BRAIN CULTURE
Shaping policy through neuroscience

Jessica Pykett

First published in Great Britain in 2017 by

Policy Press
University of Bristol
1-9 Old Park Hill
Bristol
BS2 8BB
UK
t: +44 (0)117 954 5940
pp-info@bristol.ac.uk
www.policypress.co.uk

North America office:
Policy Press
c/o The University of Chicago Press
1427 East 60th Street
Chicago, IL 60637, USA
t: +1 773 702 7700
f: +1 773 702 9756
sales@press.uchicago.edu
www.press.uchicago.edu

British Library Cataloguing in Publication Data
A catalogue record for this book is available from the British Library

Library of Congress Cataloging-in-Publication Data
A catalog record for this book has been requested

ISBN 978 1 44731 404 2 hardcover
ISBN 978-1-4473-1405-9 paperback
ISBN 978-1-4473-2146-0 ePub
ISBN 978-1-4473-2147-7 Mobi
ISBN 978-1-4473-1406-6 ePdf

Cover design by Marisa Harlington
Front cover image: http://commons.wikimedia.org/wiki/
Printed and bound in Great Britain by CPI Group (UK) Ltd, Croydon, CR0 4YY
Policy Press uses environmentally responsible print partners

For David and Eva

Contents

About the author

Jessica Pykett is a social and political geographer at the University of Birmingham, UK. Her research to date has focused on the geographies of citizenship, education and behavioural forms of governance. Her previous books include *The pedagogical state* (Routledge) and *Changing behaviours*, with Rhys Jones and Mark Whitehead (Edward Elgar). She teaches on the spatial politics of welfare, work and wealth.

Acknowledgments

This book would not have been possible without the excellent research assistance of Dr Bryony Enright and Tom Disney, supported by funding from the School of Geography, Earth and Environmental Sciences at the University of Birmingham (2012-14). I would like to thank all the interviewees who participated in the research and who shared their time and their thoughts so generously.

Ten neuroscientists and behavioural scientists based at universities across the UK also gave crucial insight into the remit, practice and significance of their work at the initial stages of the research. The ideas presented here have also benefited from several conference sessions and seminars at which geographers, psychologists, educationalists, neuroarchitects, designers, political scientists and others have helped me to develop the theoretical framework of the book.

I thank the panel at the Royal Geographical Society and Institute for British Geographers Annual Conference 2013 on 'Understanding the psycho-spatial' (Liz Bondi, Gail Davies, Peter Kraftl, Steve Pile and Mark Whitehead), and my co-organiser, Elizabeth Gagen, for contributing their brilliant thoughts at this forum. At a seminar on 'Bio-social methods for a vitalist social science' in July 2013, Felicity Callard, Megan Clinch, John Cromby, Des Fitzgerald, Kathryn Ecclestone, Martyn Hammersley, Steve Hinchliffe and Rachel Lilley provided invaluable insight into the changing terrain of the social sciences.

I greatly appreciated support from the University of Birmingham's Institute of Advanced Studies, particularly Sue Gilligan and Sarah Jeffery for enabling that seminar to happen. I also gratefully acknowledge the support of the Economic and Social Research Council (ESRC) for funding a seminar series on 'Behaviour change and psychological governance' (Grant Ref: ES/L000296/1), and the participants and speakers at these seminars. These seminars have shaped the ideas presented here.

Providing valuable additional insight and commentary on these seminars were Colin Lorne and Stacey Smith. Thanks to my co-organisers Ben Anderson, J.D. Dewsbury, Maria Fannin, Rhys Jones, Joe Painter and Mark Whitehead for their readiness to explore these themes together. Rhys and Mark in particular deserve special thanks for the opportunities they have given me to conduct research in this area, and in the many enjoyable conversations we have enjoyed at Aberystwyth, a most special university by the sea.

At Birmingham I am grateful for the support of my excellent colleagues at the School of Geography, Earth and Environmental Sciences, as well as Lindsey Appleyard, Will Leggett and Catherine Needham.

I am particularly indebted to Clive Barnett, Paul Cloke, Wendy Larner and John Morgan for teaching me about geography, and to John Clarke and Janet Newman for teaching me about social policy. I also thank Laura Vickers and Emily Watt at Policy Press, copyeditor Dawn Rushen, indexer Hilary Faulkner and the anonymous reviewer for their detailed reading of the full typescript and very helpful and generous suggestions.

Finally I would like to thank my family: Lyn and Andrew, Rachel, Kelly, Ben, Verity, Esme and Martha, and David, Iris, Idris and Eva.

Preface

The European Commission's €1.2 billion Human Brain Project and President Barack Obama's BRAIN initiative attest to contemporary governmental ambitions to uncover the mysteries of the brain, to break through new frontiers of knowledge, and to determine the brain's impact on human behaviour. These efforts to map the brain, to capitalise on the vast datasets emerging from contemporary neuroscience, and eventually to develop the computing power to simulate neural functioning are the latest indications of a culture in which the brain is privileged in its explanatory power for all manner of human experiences, decisions, capabilities, actions and relationships.

This book is about the effects of this brain culture on the governance of citizens in the UK. It outlines several social spheres in which the neurosciences, psychological and behavioural sciences inform policy and practice. It examines academic disciplines and modes of understanding which have themselves been informed by a 'neural turn'. While brain culture is not new, the book considers its current manifestations. It explores the intersections of political, economic and social practice with specific brain claims and brain-based activities. The central assertion is that this brain culture can only be understood in a specific geo-historical context – the context of the brain world. There is a circularity inherent in this assertion which poses a significant challenge to both neuroscientific attempts to know the biophysical brain, and to social scientific, arts and humanities research that sets out to comprehend brain culture.

Human geography is one disciplinary approach that should enable us to better understand the brain world, since it has long been concerned with the relationship between humans and the world, situated behaviours and spatial context. Arguably impeded by its own neural turn, and by technocratic and instrumentalised forms of knowledge, there is a need to preserve a distinctly human geographical account of the 'psycho-spatial' to offer an alternative to the potential reductionism, determinism and medicalisation of brain culture. This account treats the person as having a thinking mind embedded in a particular social space and era. It approaches the drivers of behaviour as at once embodied and discursive, but challenges the privileging of scientific accounts of the brain as a body part somewhat divorced from the world. Taking a geographical approach to context in this manner allows us to explore the world beyond biology and psychology, and to better understand the social, political, economic and cultural factors that shape behaviour *in situ*. It reveals the partiality of neuroscientific, psychological and behavioural economic description and prescription, and expands our conception of what matters in explaining why we act in particular ways.

The book interrogates the actual policies, practices and cultural formations through which the self is constituted as subject to its brain processes, through urban design, education and work training. These spheres are aimed at producing the city dweller whose attentions are managed, the biologically rendered normal learner, and the worker who emotionally labours in the pursuit of self-optimisation. By engaging with individuals and organisations working within these spheres, straightforward criticisms of brain culture as a malign manipulative force through which powerful interests control our brains and behaviours are rendered embarrassing and problematic. There are diverse actors who are involved in constructing the brain world for many different reasons and towards distinct, sometimes competing, and often laudable, goals. But neuroscientists, too, have expressed unease and mistrust at a number of representations and applications of their work. They have also openly challenged research programmes such as the Human Brain Project as nothing more than a massive scale IT project (Sample, 2014). Engaging with these individuals for the research on which this book is based has encouraged me to try to outline the specific and cumulative impacts and unintended consequences of brain culture, as well as the ways in which such a culture is actively resisted and challenged in everyday practice and public debate.

Mapping the brain world could not be more different from mapping the brain. It is a brain geography more concerned with contextualisation than with localisation, with connections on the ground rather than connectomes in the brain. In identifying the high political, philosophical and social stakes of brain culture, a lack of neuroscientific training may seem like a stumbling block. However, it is the contention of this book that the production of expertise in general needs to be explored and questioned. The book does not therefore challenge the validity, reliability and veracity of those working in neuroscience, but rather examines the cultural imagery of the brain as used in particular arenas of research, policy and practice. In so doing it hopes to cultivate a sense of hesitation in the pursuit and application of brain-based explanation, behavioural insight and psychological evidence. Such rationalities for action have a broader impact on governance and citizenship which non-expert publics must be supported to discern, deliberate and dispute.

Jessica Pykett
Birmingham, 2015

ONE

Introduction: governing through brain culture

Are you for or against 'brain culture'? Today there is extensive debate on the cultural, social and political implications of numerous insights from neuroscientific, behavioural and psychological research, from arguments over the consequences of brain-based teaching models in schools, the commercial adoption of neuroscientific methods and technologies, to the ethics of using behavioural insights to inform public policy. There are concerns around about the scientific venture of neuroscience itself, as it seemingly transforms our understandings of the human condition, everyday behaviour and social relations. This debate has been voiced in popular literatures as well as in academic studies from a wide range of disciplines including anthropology, communications and media studies, gender studies, law, sociology, psychology, philosophy, politics, the biological and medical sciences and the emerging field of critical neuroscience. The aim of this book is to situate these theoretical and cultural critiques in a more specific analysis of contemporary public policies and social practices which have begun to absorb the influence of what has been dubbed 'brain culture' (M. Taylor, 2011; Thornton, 2011). The idea of brain culture refers to the way in which knowledge, images and representations of the brain shape our cultural identities and societies. This book examines what the influence of this broad phenomenon of brain culture means for re-shaping practices of governance and citizenship in the UK.

What evidence do we have of the contemporary emergence of brain culture? As we venture further into the 21st century, more than 20 years after US President George Bush welcomed in the 1990s as the 'decade of the brain', the UK has been recognised as a social laboratory for innovations in public policy and practice which look to the behavioural and neurosciences for their rationale. In 2007, a major UK educational research initiative reported on the findings of recent decades of investigation relevant to shaping the future of neuroscience in education (Howard-Jones, 2007). The UK has also incubated a growing network of expertise in mindfulness-based cognitive therapies (MBCT) that are put to use in both mental health and workplace settings in order to relieve stress and to cultivate new

1

brain habits. Many mindfulness initiatives now look to brain imaging research techniques for their validation, offering convincing arguments for governmental support. Psychological forms of knowledge are also being used in the governance of happiness across whole populations, through statistical initiatives to measure and improve gross national happiness (GNH). In the fields of youth justice, anti-social behaviour and welfare reform, neuroscientific evidence has been used to both support and contest various forms of government intervention. For instance, in 2011, the UK's principal scientific academy, The Royal Society, published a report questioning the basis of the current age of criminal responsibility (age 10) in light of the neuroscientific evidence that shows that the brain is developmentally immature at that age (The Royal Society, 2011a). In an earlier report on the importance of early interventions in child welfare, Labour MP Graham Allen and Iain Duncan Smith (2008) reviewed the implications of the brain sciences and psychological theories of development and attachment for early child development (ages 0-3), good parenting and future citizenship. By 2013, the Early Intervention Foundation had been launched as a new think tank for early intervention, providing advice, evidence and expertise on 'what works' in improving children's emotional and social skills, in order not only to improve community welfare and reduce government spending, but also to intervene *psychologically* to 'break the intergenerational cycles of dysfunction.'[1]

What marks out UK public policy as an apparent hotbed of brain culture is the degree to which contemporary policy experimentation has been shaped explicitly by behavioural, psychological and neuroscientific evidence. Central to this venture has been the work of the Institute for Government, an independent think tank providing training and development for civil servants, and the UK Cabinet Office. They have championed behavioural science approaches to policy-making, and The Behavioural Insights Team was set up in 2010 in order to cement this approach across government. By 2013, the 'Nudge Unit', as it is also known (referring to the popular behavioural economics book from which the nudge approach has been derived; see Thaler and Sunstein, 2008), had begun exporting its expertise in the behavioural sciences to foreign governments, and now exists as a 'social purpose company' in partnership with Nesta and the Cabinet Office[2] – heralding a new

---

[1]   www.eif.org.uk

[2]   www.behaviouralinsights.co.uk/about-us

direction for strategic policy development which was previously the preserve of democratic governments.

The influence of brain culture is by no means limited to the UK. As The Behavioural Insights Team was being established in the UK, in France, the Prime Minister's advisory body, the Centre d'analyse stratégique, was beginning a programme of work on neuroscience and public policy in 2010, and has reported on the public health implications of neuroscience, psychology and behavioural economics (Oullier and Sauneron, 2010) and the legal implications of neuroscience, or 'neurodroit' (Oullier, 2012). In Australia, the UK's Nudge Unit has begun directly advising the government of New South Wales on implementing nudges.[3] Elsewhere, the Netherlands School of Public Administration (Nederlandse School voor Openbaar Bestuur, NSOB), which, like the UK's Institute for Government, provides training for civil servants as well as research and policy advice, has suggested that the Dutch government should develop its own 'Nudge Unit' based on applying the behavioural sciences to policy-making (van Oorschot et al, 2013). In Denmark, a non-profit organisation, iNUdgeYou, which designs nudges for public policy and commercial companies, has been established. In Singapore, the Ministry of Manpower has recently set up a Behavioural Insights and Design Unit, again drawing on a consultancy arrangement from the UK's Behavioural Insights Team. In the US, where much of the academic research in behavioural science has been developed, the White House has now set up a Social and Behavioral Sciences Initiative (dubbed the 'Nudge Squad') led by Maya Shankar, a cognitive neuroscientist.

Meanwhile, global bodies such as the European Union (EU), World Bank and World Economic Forum are also delving into brain-based policy discussions. In 2013 the European Commission published a guide for EU policy-makers on how to design, implement and monitor policies based on behavioural insights, and a more realistic account of the non-rational aspects of people's behaviour (van Bavell et al, 2013). The World Economic Forum set up a Global Agenda Council on Neuroscience and Behaviour (2012-14) to bring together insights from neuroscience and psychology to promote behaviour change policies globally. And economists at The World Bank are dedicating the 2015 *World development report* to understanding the behavioural and social foundations of development, seeking insights

---

[3]   www.smh.com.au/comment/what-a-difference-a-nudge-in-the-right-direction-can-make-20130406-2hdab.html

from: behavioural economics; social, cognitive and cultural psychology; sociology; anthropology, and especially cognitive anthropology; and neuroscience.[4] Such examples are no doubt only in their infancy in terms of making policy interventions with widespread effects on the ground. However, they indicate the experimental endeavours of a number of national governments, non-governmental organisations (NGOs) and global institutions in attempting to re-imagine policy-making in an era of brain culture. This is an era in which the cognitive, emotional, neurobiological and behavioural processes of the citizen are seen as the new target points of strategic, intelligent and effective policy strategy.

These policy experiments in brain culture have been shaped to a large degree by research developments in the US, where in recent years, research institutes for the applied 'decision sciences' and behavioural sciences have grown in significance. To give a few examples, Richard Thaler, a behavioural economist and co-author of *Nudge*, directs the Center for Decision Research at the University of Chicago in addition to having provided advice to the UK and French governments. Duke University in North Carolina has recently set up the Behavioral Science and Policy Center, and UCLA's School of Management formed an Interdisciplinary Group in Behavioural Decision Making in 2003. Harvard University's Foundations of Human Behavior Initiative led by David Laibson is well known for its Russell Sage Foundation summer schools in behavioural economics and neuroeconomics.

It is also in the US that the fields of neuroarchitecture, neuroeducation and positive psychology have been initiated. Private philanthropic organisations such as The Dana Foundation, based in New York, have led the way in providing grants for neuroscientific research, coordinating an annual 'Brain Awareness Week', providing resources for the popular understanding of the brain sciences and developing approaches to learning-based neuroscientific developments. Similarly, the John Templeton Foundation has supported research, institutions, networks and prizes in positive psychology and positive neuroscience. Meanwhile the Academy for Neuroscience in Architecture (ANFA) was set up in San Diego in 2003, in order to bring together neuroscientists and architects to forward knowledge of human responses to the built environment.

---

[4]   www.sciencespo.fr/newsletter/actu_medias/4086/Notice%20of%20Job%20
Opening%20for%20WDR%20%20June%202013_July%202014.pdf

Major research initiatives for the neurosciences have recently been launched. In April 2013, US President Barack Obama launched a new programme of research, and pledged US$100 million of initial investment (from state, private, philanthropic and university sources) for the BRAIN initiative (brain research through advancing innovative neurotechnologies), set to establish the 2010s as the new decade of the brain. Involving the National Science Foundation (NSF), National Institutes of Health (NIH) and the Defence Advanced Research Projects Agency (DARPA), this wide-ranging programme brings together the US military, health research and scientific communities in a new strategic alliance. Some of the goals of the BRAIN initiative include understanding 'how brain activity leads to perception, decision making and ultimately action', and producing 'a sophisticated understanding of the brain, from individual genes to neuronal circuits to behaviour.'[5] Similarly, the Human Brain Project funded by the European Commission's Future and Emerging Technologies initiative (2013-16) has set out to build new computing and medical capacity which might help to develop medical databases of and treatments for the brain, technological replication and emulation of the brain, and to shape our understandings of what it means to be human.[6]

Such globally ambitious goals for investment in and understandings of the brain are suggestive of how the brain is increasingly seen as a fundamental research frontier in the quest to understand what lies behind human behaviour and action. This poses a significant challenge to philosophical, ethical, political and legal debates around the notion of free will, responsibility and indeed, personhood. Yet the Human Brain Project, for instance, has reserved only 5 per cent of its €1.2 billion budget for examining the implications of these radical advances for humanity. It should be noted that such debates are not the preserve of philosophers, political scientists or academics in general. The place of the brain in our cultures, our politics and our identities has garnered wide popular interest. In 2011, Matthew Taylor, chief executive of the Royal Society for the Arts, Manufacturing and Commerce (RSA) think tank, and former policy adviser to Tony Blair, explored some of these issues in a series of radio programmes aimed at investigating the coming of 'brain culture'.[7] His programmes outlined how the neurosciences and brain scanning in particular have begun to change

[5] www.whitehouse.gov/blog/2013/04/02/brain-initiative-challenges-researchers-unlock-mysteries-human-mind
[6] www.humanbrainproject.eu/en_GB
[7] www.bbc.co.uk/programmes/b017n523/episodes/guide

the world around us, including our institutions, social attitudes and conceptions of self. By looking in particular at the implications of neuroscience for criminal law, educational practice and public policies based on behaviour change, he raised crucial questions of how we might both make better choices and avoid manipulation within this new brain culture.

In the same year, communications scholar Davi Johnson Thornton published a book also entitled *Brain culture* (2011), which examined the brain as a rhetorical phenomenon in popular culture. She provided an original analysis of the circulation of the very concept of the brain in popular media, public discourse and in practice, showing how the brain is mediated by social contexts. She explored the way in which the 'idea' of the brain is mediated – primarily through brain scanning images and media reports and popular assertions about the brain as self – and that this has real social effects. Thornton's aim was to understand the context in which neuroscience informs our perceptions and explanations of reality, causation and truth:

> I try not to make it sound as if there is some true story of the brain that scientists and journalists have mistranslated or misunderstood. As a scholar of public discourse, I am generally less interested in whether it's true or false, and more interested in what kind of effects it has in particular cultural contexts.[8]

This position is helpful in framing the approach of *Brain culture* which, like Thornton's work, is interested in the actual social and political effects of the neuroscientific, psychological and behavioural imaginary on the ground, in particular historical and geographical contexts. It thus takes the relationship between the brain and the world as its central focus. Thornton's field of interest relates specifically to popular texts and representations of the brain, mainly in the US, including not only brain scan images, but also, among other things, self-help literature, video games and *Time* magazine features on early childhood brain development. By contrast, the aim of this book is to understand how specific sites of public policy and practice have both been shaped by and shape brain culture. In so doing, the book examines the consequences of these practices for changing cultures of governance and re-inventing citizenship.

---

[8] www.wired.com/2011/08/the-rhetoric-of-neuroscience/

The book is interested in the specific effects of a culture which has been shaped by neuroscientific, behavioural and psychological research insights, and examines these effects within particular fields of practice: neuroarchitecture, neuroeducation and positive psychology in the workplace. In this respect, the book is indebted to a growing cadre of critical scholars of the psychological and neurosciences who take the French philosopher Michel Foucault's notions of governmentality and biopolitics, or the politics of 'life itself', as their analytical starting points. Foucault's influence has arguably been essential in contemporary readings of the cultural, social, economic and political significance of the 'brain world'. Tracing the relationships between knowledge, power and human subjectivity has been crucial in navigating through critical debates that cast the neurosciences either as the core of or a fundamental threat to human agency. This thread is found not least in Thornton's own account of popular neuroscience as a governmental technology, and in her analysis of the key paradox of its person-shaping capacities. The rhetorical brain of popular neuroscience, she argues, is at once claimed as the ultimate driver of your behaviour, and as a resource to be shaped by you. The brain can be moulded and worked on in order to improve its function, its resilience, its health – partaking in 'endless projects of self-optimization in which individuals are responsible for continuously working on their own brains to produce themselves as better parents, workers and citizens' (Thornton, 2011, p 2).

This work suggests that there is more to behavioural, psychological and neuroscientific endeavour than 'straightforward' (although it is far from this) scientific investigation. But before we turn to some of the specific effects of brain culture on our ideas and practices of person-shaping, citizenship and governance, it is worthwhile examining some of the limitations of neuroscience itself. The remainder of this chapter provides an introductory tour of what might be called 'neuroskeptic' literatures, giving an indication of just how influential the neural turn has been in a wide range of academic disciplines hitherto only analysed separately. It then outlines the basis for the critical analysis of the brain world adopted in the book by exploring the role of cultures of the brain in shaping citizen subjectivity through public policy.

Introducing the neuroskeptics

This book is concerned with the effects of culturally mediated knowledge about the brain, the psyche and behaviour on policy and practice, as distinct from neuroscience itself. Neuroscience is not an internally coherent totality – there are a range of approaches, several different methods, different topics of study, contrasting models of neural processes, and significant debates concerning controversial phenomena such as consciousness, intentionality and causation. We must therefore be wary of clumping together the whole neuroscientific endeavour, let alone the psychological and behavioural sciences that also feature prominently in this book. It is important to acknowledge the distinct trajectories of these disciplines. While mathematics and game theory have been crucial in shaping behavioural economics, the impact of biology, chemistry, physics and technological development on the neurosciences has been notable, while there has long been significant debate on the place of psychology within the social or life sciences. Nonetheless, it is still helpful to discuss quite specifically the unintended consequences of *brain culture*. The term 'brain culture' describes how neuroscientific, psychological and behavioural accounts of human beings (and the scientific authority associated with them) are increasingly shaping our relationship to ourselves, the world and to each other, in ways that are manifest in policy and practice. It is therefore essential to question the potential assumptions of and partial accounts offered within the brain and behavioural sciences, and to highlight some of the scientific uncertainties that persist within these fields. While this book focuses primarily on actually existing brain culture in specific social worlds, it is certainly worth considering the foundations on which contemporary neuroscience as a substantial corpus of expertise is built.

While it is not necessary to be a neuroscientist to offer an effective analysis, it is useful to give due regard to the critiques of neuroscience offered by neuroscientists and psychologists of their own fields of enquiry. Two such critics, Kathleen Taylor and Raymond Tallis, both with past careers as neuroscientists themselves (Taylor as a cognitive neuroscientist, and Tallis as a clinical neuroscientist) are among a group of authors whose recent aim has been to draw the attention of the UK public to the potential consequences of *The brain supremacy* (Taylor, 2012) and 'neuromania' (Tallis, 2011). Both authors recognise that important clinical advances in the neurosciences have had tangible effects on patient health and the lives of those suffering in particular with neurological disorders or brain traumas. Neuroscientists such as Taylor, who take a role in shaping public understandings of neuroscience, warn that the ethical perils of neuroscience (specifically, the brain's potential vulnerability to manipulation) equal their potential. In this sense, there is a pressing need to set out legislative frameworks to stave off the possible abuses of neuroscientific knowledge and to uncover the vested interests implicated in neuroscience research and applications. In other words, it has become essential to investigate the contexts in which neuroscience is produced, used and communicated in brain policy, practice and culture.

Even more damning of the neuroscientific venture is Raymond Tallis (2011), who, in *Aping mankind*, marks the proliferating applications of neuroscience (the likes of neuroliterary criticism, neuroeconomics, neurolaw, neuroethics and neurotheology) as fundamental threats to our understandings of humanity. Such research programmes, cultural formations, policies and practices render us subservient to our pre-historical, animal selves (indeed, the self, too, is eroded by this 'neuromania'). He helpfully outlines both the methodological shortcomings and flawed philosophical logics of a significant tranche of neuroscientific research, leaving little of the neuroscientific 'project' unscathed. The first limitation relates to the technological measurement of brain activity itself. The fMRI (functional magnetic resource imaging) scans which purport to give their viewers, media commentators and many neuroscientists a 'window' onto the brain, imbued with so much authority and visual power, are themselves only proximate interpretations of brain activity. Not only do they not actually show 'brain function' per se, but also rather a proxy measure of blood flow, and their broad spatial and temporal resolution makes them somewhat clumsy approximations of actual neural activity. Kathleen Taylor brought the magnitude of the situation to light in a talk given to the public at the Hay Festival in 2013. As she explained to a large audience apparently concerned by the advent of brain culture, neuroscience is a science in its infancy. The number of neurons in the brain, she recounted, amounts to 24 times the world's population, with no existing computational power able to cope with analysing such figures. Taylor also recollected that when fMRI was first introduced, many cognitive scientists dismissed the new technologies as nothing more than mere 'brain geography' (Taylor, 2012, p 34), a descriptive tool to map various functions, but which offered little of note to the advancement of existing concepts of the mind and theories of cognitive processes. Yet since the inception of fMRI, exploration of the brain, the drive to push back the frontiers of brain science, and the search for the brain centres and processes associated with our fundamental behaviours and emotions have not ceased. The basic neuroscientific endeavour thus remains infused with an implicit geographical imaginary founded in a colonial frontier vision of geographical exploration and modern progress. This conception has been significantly destabilised by advances in the discipline of human geography as it is practised today, as the next chapter outlines.

The evocation of a mere 'brain geography' brings us to the second methodological limitation of the contemporary neurosciences. For many detractors, the widespread use of fMRI, its prominent depiction in media reports of new discoveries relating to the brain, and its imperative to 'localise' this or that brain function, is reminiscent of the now discredited Victorian science of phrenology. Developed by early 19th-century German physicians such as Franz Joseph Gall and Johann Gaspar Spurzheim, phrenology asserted that a person's psychology and moral character could be ascertained by an examination of the bumps and contours of their skull. Phrenology was associated in particular with early 20th-century investigations into the criminal mind, and is recognised as an important, if flawed, precursor to the development of present-day forensic neuroscience (Davies and Beech, 2012). But even to its contemporaries, phrenology was a highly contested pursuit, and its spread around Europe and America was met with scientific contestation. It was therefore a relatively short-lived enterprise, associated with 'quack' doctors, and dismissed as a pseudo-science. Yet for critics, the authoritative contemporary neurosciences harbour a lingering gesture towards the pseudo-science of phrenology, that is, the localisation of specific brain functions and the rush to 'map' neural activity.

Psychologist William Uttal (2001) has been one critic of what he terms 'the new phrenology', accusing neuropsychologists of a quite basic fundamental error, which is to fail to adequately define the complex *psychological* processes in which certain regions of the brain are said to specialise. He does not deny that brain functions can be associated with certain areas, but that there is no scientific means by which these can be correlated with mental processes, let alone actual observable behaviours (Uttal, 2001, p 15). Not only are many mental processes without unequivocally shared definitions, but the localisation thesis is also hampered by its explicit ignorance of the role of the whole brain in performing even the apparently simplest of functions.

The third principal methodological objection to neuroscience that has been raised by several critics is that lab-based experiments have little transferable currency into the real world. In particular, there is a propensity, and one might say, necessity, for neuroscientists to reduce complex behaviour, phenomena, emotions, personality traits, moral dispositions, characterological vices and virtues to simple stimulus–response models, so that they may be investigated in experimental settings and picked up by brain scanners. The 'puzzlingly high correlations' between brain activation as measured by fMRI scanning and the personality traits they are supposed to measure have not gone unnoticed (Vul et al, 2009). For many critics, it is a circularity of logic used to design and undertake experiments which accounts for the relative success of neuroscientists to find just what they are looking for. This is, in part, as Tallis argues, because they have reduced what we might understand as an experiential phenomena (such as love, wisdom, beauty) to a specific task which might be said to stimulate a loving, wise, appreciative reaction, and then to look at the particular neural activity of the brain which corresponds with the task. Anthropologists and sociologists of neuroscience have raised similar questions. By observing lab practice and preparatory meetings to set up a neuroscientific investigation into the phenomenon of 'disgust', anthropologist Simon Cohn notes that the neuroscientific method must necessarily reduce the conscious experience of disgust to a basic mental state that can be identified in the context of the scanning technologies available. In this sense, it is not really the social emotion of disgust, with all its social baggage, the influence of a person's upbringing, its cultural associations, and a conscious person's actual experience of it that the experimenter is interested in. All these issues are 'factored out', as Cohn puts it (2011, p 186), in the interest of identifying a localised response to a specific stimulus (in this case, a series of photos meant to depict 'disgusting' things).

In this sense, the act of locating brain processes can lead to a curious *dislocation* of the psyche from its context, including everything that gives the mind, our experiences, characters and interpretations meaning. One of the central aims of this book, by contrast, is to re-contextualise brain culture through a new reading and re-scaling of 'brain geography'. A 'psycho-spatial' approach is developed in order to shed light on how the practices and applications of brain culture shape how we see ourselves, our interactions with the world and each other in real spatial contexts. In this way, the book is clearly situated within the academic discipline of human geography, explored in depth in Chapter Two. The book therefore begins to address the persistent tendency of the contemporary neurosciences to underplay this wider scale of geography by approaching the 'brain in a vat', captured eloquently by Tallis (2011, p 237):

> So to try to find our public spaces, lit with explicitness, in the private intracranial darkness of the organism illuminated by fMRI scans and other technology is to look right past what makes us human beings, and makes us what we, and our lives, are.

While the methodological limitations outlined here may be enough to cast doubt over some of the authoritative claims made in the name of the neurosciences within what has come to be known as brain culture, there are further philosophical objections which may also be held up as a challenge to the increasing explanatory power of neuroscience in many diverse fields of policy and practice. One might remark that there is good and bad science in every domain, not least those domains which are subject to political debate. But if there is an even more substantial problem with brain culture, it is that it is philosophically and politically wedded to the notion that science will tell us how to live better, wealthier, healthier and happier. Many contemporary critics – clinical neurologists, critical psychologists philosophers, anthropologists, sociologists and political theorists alike – are highly sceptical of the claim that brain functions shape our societies and identities, let alone that knowledge of these brain functions *should* be used in developing our public policies and everyday practices.

Some of the more philosophical concerns are outlined again by Tallis, from the perspective of philosophies of mind – of which I only summarise three aspects here as they relate to philosophies of *situated* minds. First is the tendency of neuroscientists to equate correlation with both causation and identity. Evidence in a brain scan is neither proof of a causal relationship between stimulus and corresponding brain activity, nor is it identical to what people actually experience (Tallis, 2011, p 87). Indeed, it can be said that neuroscientists mistrust conscious experience – why ask people what they are thinking and doing if the majority of human action is beyond our conscious awareness? The principal oversight is that the biophysical materiality of the brain can only ever be discerned as a (neuroscientific) representation of any given reality or experience *in and of the world*. In this way, neuroscience can only exist in the presence of a conscious observer, an observer of which neuroscience wishes to deny the very existence. Second, human perception cannot be reduced to neural processes because to do so is to suggest this is a one-way street. Rather, as Tallis (2011, p 109) explains, 'human beings are not simply organisms but rather are *embodied subjects*', by which he means that humans do not simply respond to the world as if it were only an ecological system. Rather, we perceive as knowing observers who are at once materially embedded in and consciously distant from the world. Our consciousness, in these terms, always has to be *about* something. His assertion is that it is not functional brains that see and perceive, but people with minds thinking about real things in the world. Related to this, Tallis's third problem with 'neuromania' is that it proposes a personless view from nowhere and a denial of the situated self. In positing a materialist and objective account of the brain as an organ within an organism, neuroscience undermines 'the sense of a centred world, of "me", or of the ownership that makes a brain *my* brain, *a* body, *my* body, a portion of matter *my* world' (Tallis, 2011, p 114, original emphasis).

Several other philosophers of mind share Tallis's concerns about the omnipresent explanatory power of the neurosciences. Those who, like Susan Hurley and Alva Noë, regard brain culture – the enthusiasm surrounding brain imaging, neuroscientific advances and real-world applications – to offer misleading accounts of consciousness. This enthusiasm leads to dogmatic claims that neuroscience will eventually provide the comprehensive answer to the intractable question of what it means to experience being human. Preferring a conception of consciousness *in action* (after Hurley, 1998), Hurley and Noë (2003, p 132), by contrast, argue that:

> To find explanations of the qualitative character of experience, our gaze should be extended outward, to the dynamic relations between brain, body, and world.

This avowedly situated approach to the philosophy of mind has challenged the computational model of the brain as a processor of information perceived from the world and translated into the output of action. It has replaced this model with the conception of consciousness as being both embedded (in the world) and enacted (by people). As Noë explains (2009, p 24), the naïve materialism of mainstream and popular neuroscience, and its celebration of the brain as the source of human activity, is entirely misplaced. Rather:

> … [c]onsciousness isn't something that happens inside us: it is something that we do, actively, in our dynamic interaction with the world around us […] if we want to understand how the brain contributes to consciousness, we need to look at the brain's job in relation to the larger nonbrain body and the environment in which we find ourselves.

The notion of a 'brain world' offered here suggests a focus on a more psycho-spatial account of consciousness, in which the mind – although fleshy, material, biochemical and physical – can only be properly understood in context. That context is a complex social terrain that is geographically and historically specific. It is further complicated by the fact that the 'environment in which we find ourselves' is one that has already been shaped by brain culture – our sense of personhood is intimately related to this very culture which claims to tell us so much about ourselves.

How does brain culture make people?

To be troubled by philosophers of mind may seem a purely academic concern. I have sketched out some of their concerns here because of the a priori challenge that they pose for mainstream neuroscience and our common-sense understandings of human biology and physiology. And yet there are also very real outcomes where neurophysiological accounts of the mind pervade our culture, shaping how we feel, think and act. Psychologist Cordelia Fine (2010), for instance, reveals how our culturally specific conceptions of gender differences can be reinforced by spurious and misleading neuroscientific accounts of sex differences in the brain that are taken as fact by virtue of the rhetorical authority of the brain scan. For Fine, there is a certain 'neurosexism' at work, which can shape personal relationships, the status of men and women in public life, work practices, and the education of children.

Several other authors have showed how brain culture can shape activities, in business, commerce, schooling, family life, law, military training and mental health, for instance. But can brain culture also shape our very sense of self? For many sociologists, philosophers and anthropologists, this is precisely the case in the production of 'brainhood' (Vidal, 2009), 'cerebral subjects' (Ortega, 2009) or 'neurochemical selves' (Rose, 2003). Such authors point to the way in which the evolution of the psychological and now the neurosciences (and genetic sciences) have yielded a new cultural affinity with the biological organ that is the brain. We are coming to see ourselves, Rose (2007, p 96) argues, as 'somatic individuals', thinking of ourselves as embodied 'in the language of contemporary biomedicine'. For Fernando Vidal, of the Max Planck Institute for the History of Science in Berlin (a centre for critical thought on the history and philosophy of the neurosciences), the cerebral subject refers to 'the property or quality of *being*, rather than simply *having*, a brain' (Vidal, 2009, p 6, original emphasis). One's personhood has for many neuroscientists become identical to one's brain, encapsulated by the often-quoted phrase 'you are your brain' (a sentiment shared by neuroscientist Micheal Gazzaniga, Nobel prize-winning neurobiologist Francis Crick, philosopher Daniel Dennett, and neurophilosopher Patricia Churchland, among others). But crucially, this brainhood is not a radical break with the past, but rather, as Vidal notes, a continuation of modernist conceptions of individuality and the interiority of the self. It is not therefore the case that brain culture simply 'makes people' by shaping how we view our sense of agency, personality and identity. Instead, our dominant *cultural* conceptions of the modern self permit the kinds of assumptions that neuroscientists make about personhood in the course of their investigations, which, in turn, reproduces the 'supreme [Western] value given to the individual as autonomous agent of choice and initiative' (Vidal, 2009, p 7).

For scholars such as Vidal, brain culture is not produced through scientific discourse; our cultural representations shape that very science. In this sense, neuroscience is not simply the driving force behind a brain world, but is formulated in the context of that brain world. However, as Pickersgill et al (2011) have warned, it is important for anthropologists, historians and sociologists of neuroscience to avoid over-emphasising the existence of brainhood or brain culture. In their study of the views of patient groups, neuroscientists, teachers, foster care workers and clergy on the topic of neuroscience and the self, Pickersgill and colleagues found that while their respondents did express 'new' understandings of themselves and others based in a language of neuroscience, they also drew on much longer-standing conceptions of self and society. Their self-narratives regarding the brain were highly context-specific, and there was much resistance to the idea that the brain was the sole source of identity. Sometimes it was seen as 'just another body part' (2011, p 362).

While many appreciate that neuroscience is in its infancy and application of its findings fraught with uncertainties, or indeed that its findings are relatively modest, or specific to a very small number of tasks and contexts, this book is concerned with the cumulative effect of a more generic notion of brain culture. It does not therefore engage in great detail with original neuroscience research published in its peer-reviewed journals. Instead, its reference point is the way in which the psychological, behavioural and neurosciences are together increasingly viewed as the last word (in diverse spheres of research, policy and practice) on the drivers behind what makes us people, and by extension, citizens. The repercussions of such accounts of our behaviour in different social fields of practice are explored in each of the following chapters.

The specific effects of brain culture for citizens are already evident in a number of spheres, not least within the aforementioned domain of policy experimentation gripping the UK's civil service, Whitehall departments and influential London think tanks. Most notably, the RSA think tank has spearheaded a campaign for 'neurological reflexivity', understood as cultivating an ability to shape and reshape our brains. Its director, Matthew Taylor (2007), has argued that the challenges facing humanity *as a species* require a new sense of selfhood based on new scientific discoveries. These discoveries, he has asserted, include the fact that our sense of consciousness is an illusion made up of automatic mental processes. This involves the abolition of the 'I' inside my head, which, 'the science tells us' is driven by the electric impulse and chemical determinants which control activity in our more conscious areas of the brain. Furthermore, we are 'hard-wired' to make cognitive errors, our decisions are often self-defeating, and our thinking is often unreliable (assertions as much at home in neuroscience as within behavioural science). More fundamentally even, neural activity is said to precede conscious thought, proving that even when we think we are thinking, we are, in fact, already acting automatically. This last assertion, derived from the experiments of neuroscientist Benjamin Libet and colleagues, has been extensively debated in the field of human geography, and will be discussed in the following chapter.

Taylor went on in his speech to make the uncompromising assertion that: 'at the frontier of brain science we will continue to see a steady flow of developments which will continue to undermine our common sense idea of selfhood and our assumption about the unchanging human nature' (Taylor, 2007). The aims of his political vision for the neurologically reflexive person may be attractive: to dismantle the individualist, consumer capitalist and environmentally destructive modern self; to pay attention to our connectivity to the material world, the contexts in which our self is shaped; and to build strong civic institutions which recognise our essential capacities for empathy, altruism and collective action. Yet his appraisal of the citizen as an irrational actor evacuated of free will (and yet responsible for the self-improvement denoted by brain plasticity) is a troubling one for politics in general.

Political scientists, too, have not been shy in taking up the neuroscientific challenge. Novel conceptions of democracy, political decision-making and political agency have been promoted via neuroscientific and psychological insights. Perhaps most well-known is the work of Neuman and colleagues, who contend that affective neuroscience must necessarily transform political science, particularly its reliance on models of rational choice and deliberative judgement (Neuman et al, 2007, p 7):

Concerns about citizen irrationality are intimately intertwined with judgements about democratic practice itself. If citizens are easily distracted in their judgements, it is better to turn to a philosopher king, or at least an elite-oriented form of representative government that insulates policy from the undulating passions of the madding crowd.

The justification of the rule of a cadre of experts over an unpredictable and emotionally charged majority is not particularly new, but it is a no less dangerous conception. Elsewhere, it has been claimed that a 'political geography of affective intelligence' (MacKuen et al, 2007, p 126) misleads us into separating off the rational and emotional aspects of political judgement. This spatial metaphor holds that we can separate political decisions into (1) familiar situations in which we rely on heuristics, make immediate, automatic and habitual decisions, and (2) uncertain and unfamiliar contexts in which we cannot adequately rely on these automatic processes and must therefore reflexively evaluate and deliberate. For MacKuen et al (2007, p 150, emphasis added) this denies the role that affect plays in *all* political decision-making, which is more temporally dependent:

> The spatial metaphor misleads us by distracting us from the temporary sequencing tasks that are the core responsibilities of preconscious appraisal systems, responsibilities that control and are followed by conscious awareness and *thereafter* by introspection and reflection.

As the next chapter will argue, far from being misleading, there is an urgent political need to hold on to the spatial metaphor that reminds us of the role of *context* in driving human behaviour. Context is crucial in shaping social norms, personal identities, relationships, emotional response, embodied action and what counts as 'reasonable' behaviour. There is, of course, a significant amount of debate within political science as to the hierarchical divisions implied here between (culturally specific) reason and (biologically driven) emotion. The prioritisation of a deterministic temporal sequencing of the non-cognitive drivers of human judgement has also been much discussed, not least by feminist political theorists (Prokhovnick, 1999; Nussbaum, 2001; Krause, 2008). One recent provocation from John Hibbing (2013) called for political scientists to take a more neurobiological approach to political decision-making, and stimulated several illuminating responses. The tenor of this debate and those like it reminds us that arguments over the reductionism, determinism and medicalisation of knowledge of personhood and political agency are far from over. In her response to Hibbing, Linda Zerilli (2013), for instance, forcefully argues that a neurobiological approach to political judgement is too wedded to deterministic accounts of human consciousness to be of any use to democratic practices that are based on our capacity for freedom. She concedes that political theorists need a better conception of the body and emotion, but argues that neurobiology should not be mistaken for an account of humanity. Rather, it is an account of the material brain within human organisms, which is not the same thing at all (Zerilli, 2013, p 514). Cautioning that history is full of examples of the dangers of political scientism and technocratic approaches to governance, we are reminded of the high political stakes at work within competing accounts of personhood and citizenship.

Within the many social science and humanities disciplines sketched out here, several scholars who we might expect to pose a critical challenge to the new orthodoxy of brain culture – philosophers, sociologists, anthropologists, political theorists, human geographers and more – have, in fact, welcomed the 'neural turn'. They are very much part of the brain world described in this book. As already noted, proposals have been forwarded for a neurobiological political science (Hibbing, 2013) and a bio-social science (Rose, 2013a), and these developments are discussed in more detail in the concluding chapter. Such trends may have significant ramifications for how we conceive of personhood, what is taken as relevant knowledge within brain culture, and what counts as valid evidence for policy and practice. The next chapter considers challenges to the neural turn in human geography, and the context of human action, subjectivity and brain culture itself. It argues for a stronger focus on the adoption of this brain culture in other academic disciplines and cognate fields of practice. The specific fields of architecture, education and positive psychology in the workplace, which are given full consideration in each of the subsequent chapters. In the following section, I pause to dwell on literary scholarship, which has furnished us with fundamental insights into personhood, character, moral judgement and deliberations on humanity. What might such research have to offer a context-sensitive analysis of brain culture?

Why brain culture is not new: the Gothic brain world

Part of the popular appeal of the neurosciences, its attractiveness to journalists and headline-writers, and its promise in terms of applications and implications for human identity and social practice, is its novelty. New behavioural correlations in the brain are located, significant neural connections are discovered, new insights are revealed. Several scholars within literary criticism have also been attracted to this novelty, from analysing 'your brain on Shakespeare' to searching for the neural correlates of aesthetic taste, rhyming, puns and other literary devices. In a recent positive appraisal of this emerging field, Patrick Hogan is careful to reserve a role for literary studies to provide aesthetic judgement, ethical and political evaluation and interpretation (Hogan, 2014, p 297). As has been noted elsewhere, however, neuroliterary criticism has had a tendency to decontextualise, universalise and dehistorise such interpretations in its quest for neurobiological traces of research participants' responses to specific pages of texts (Tallis, 2011, p 298). The most fruitful contribution of these literary scholars might therefore be to resist the tendency towards neuroliterary criticism, and instead to demonstrate how the popular embrace of brain culture is both far from new and contextually specific.

A few noteworthy examples are worth outlining here, since they tell us something very interesting about the long history of the brain world. Such work outlines the changing relationship between the sciences and the arts, the role of the neurosciences in shaping debates around free will and the human soul, and the importance of context in mediating brain culture. In an examination of the relationship between the late 19th-century Gothic romantic novel and their contemporaneous brain sciences, Anne Stiles (2012) has traced specific connections between neurologists and novelists. David Ferrier, whose experiments involving applying electrical currents to the brains of live animals in order to map cortical functions, for instance, provided inspiration for novelists such as Wilkie Collins and H.G. Wells. In part, Stiles argues, such novelists referred to these experiments to evoke horror, but also to help readers reconcile themselves with the equally horrifying conclusions of cerebral localisation – that the human soul was nothing more than a figment of our mechanistic imaginations, throwing notions of moral responsibility into disarray. Brainhood was thus as much a concern for the late 19th century as it is for our contemporary context. Novels such as Bram Stoker's *Dracula* and Robert Louis Stevenson's *Strange case of Dr Jekyll and Mr Hyde* explored neuroscientific challenges to divinity, to the boundaries between humans and animals or machines, and to voluntary will. Stiles thus describes that 'late-Victorian neurology could justly be characterised as a Gothic science' (2012, p 10).

However, her argument is not that these authors simply reflected contemporary scientific discourse, but that their explorations of character and subjectivity posed a significant challenge to the biological determinism of the brain scientists. Their novels, in fact, *shaped* scientific thought and practice. One example of this was in the development of case narratives in psychological and medical research (Stiles, 2012, p 15). And unlike many facets of 21st-century popular culture, which, as Thornton and others have described, have so readily adopted the implied moral philosophy of the neurosciences, the Gothic novel testifies to a strong refusal to be subjectified as a 'cerebral subject'. This literary culture urges its readers to pay attention to experience rather than pre-destined mental function. Furthermore, as Stiles notes, there was little hint of the paradigmatic wars that exist today between the 'hard sciences' and the humanities, arts and social sciences; novelists such as Stoker and Stevenson had both received scientific training. Describing an earlier era, Alan Richardson (2001) has also remarked that the British Romantic novelists (1790s-1830s) were well integrated into scientific communities. There was not the modern split between the sciences and arts that characterised 20th-century intellectual life – Coleridge was, for a time, closely associated with the 18th-century neurologist David Hartley, and Erasmus Darwin dabbled in poetry. That said, dialogue between the Romantic novelists and these brain scientists was often far from amicable. Coleridge eventually turned against Hartley, whose theory of 'vibratiuncles' sought to oppose the dualism between mind and brain and replace it with a materialist explanation of mental phenomena and environmental perception (Richardson, 2001, p 9). This theory held that the environment was made up of 'motions' that, through the senses, caused vibrations along a solid nervous and porous 'ether', triggering smaller vibratiuncles which resided in the brain as dispositions. It thus heralded a challenge to the notion of free will, the soul and the self. The theory quite quickly became subject to ridicule, a scientific truth superseded by new developments in physiology and biology. However, there are distinct resonances in this theory with modern neuroscientific thought on neural networks (Richardson, 2001, p 11). Likewise, the widely derided pseudo-science of phrenology that again threatened the prevailing conceptions of mind, soul and the very existence of God, was explored extensively in Gothic novels. Although it, too, was dismissed for its clumsy approximations of localised brain functions, the overall imperative for cerebral localisation lives. This is despite concerns voiced not least by neuroscientists themselves (see, for example, Brett et al, 2002).

The point here is that Richardson's take on the political significance of the brain sciences of that era shows how important it is to situate the relationships between brain science and brain culture in its social, political and economic context. While today, neuroscientists, behavioural scientists and psychologists enjoy a high status in public debate as experts in diverse spheres and as advisers to governments, in the mid–19th century, Richardson (2001, p 2) notes that there was great mistrust of these 'new materialist and naturalistic models of mind in a period of revolution and reaction, when to challenge orthodox notions of the mind and soul meant implicitly to challenge the social order.' The value of literary scholarship of this nature should not be under-estimated, particularly in an academic context such as the UK, where the humanities are increasingly institutionally and financially marginalised. While the 'insights' of such research are less obviously policy-focused and 'new' than those promoted in the neurosciences, literary criticism and other interpretative schemas are crucial for challenging the naïve adoption of contemporary scientific truths in culture, policy and practice. There remains much important work to be done in order to lay out precisely what it is about the early 21st century that has shaped brain culture in quite the way that it has.

How does brain culture re-invent citizens?

The troubled notion of brainhood (for philosophers, sociologists, political scientists and literary scholars alike) has broad implications for our theories and practices of governance and citizenship. Its diminished sense of free will and its problematisation of the rational self both challenge commonplace assumptions of political agency. Conversely, the moral directive to undertake infinite work on the plastic brain for the attainment of self-perfection and mastery are as much of concern to political actors who might wish to shape new policies and political formations around this 'cerebral subject' as to political critics who seek to contest these goals. But why do these seemingly academic and abstract conceptions of personhood actually matter? How do they shape actual people in real places? Exactly *how* this idea of neurobiological personhood circulates within specific social, political, economic, geographical and historical contexts remains to be fully elaborated. Pickersgill (2013, p 332) summarises how the neurosciences have informed state practice within the military, law, education and mental health treatment – although the implications of this for citizenship, he argues, cannot be assumed. Several recent contributions to understanding the potential impacts of neuroscientific knowledge on citizenship and governance have been inspired by Foucault's theories of governmentality (Binkley, 2011a, 2011b; Thornton, 2011; Rose and Abi-Rached, 2013). In pursuing this theoretical path, we must bear in mind the warnings of Pickersgill (2013, p 323), that Foucauldian analyses of brain culture might themselves over-state its existence and its novelty, and thus falsely reinforce its status as an object of study. Yet notwithstanding these potential limitations, it remains useful (if increasingly unfashionable) to develop a critical account of contemporary brain culture informed by this broadly conceived Foucauldian approach, for three principle reasons. First, it prompts us to outline the specific processes and practices by which brain-based forms of subjectivity are shaped. Second, it helps us to properly explore the contradictions and tensions within actually existing brain culture on the ground. And finally, it evokes a concern with both the discursive rationalities and embodied experiences of the brain world. These aims are pursued in this book through an approach that goes beyond textual interpretations and engages in real dialogue with people who might promote or challenge brain culture, through qualitative research interviews. The approach is intended to shed new light on brain culture as a form of governmentality *on the ground* and *in context*. Existing governmentality-inspired analyses of brain culture have pointed towards at least five processes which are central to the re-invention of citizens through popular evocations of neuroscience: governing the self; normalisation; the cultivation of a neuromolecular gaze; governing through the brain; and subjectification. These processes are explored to the varying degrees with which they are in evidence in the fields of architecture, education and workplace training in subsequent chapters. They go some way to explaining the significance of our relationship with the brain world in both practical and philosophical terms.

Governing the self

The first process, as emphasised by Thornton, is the way in which popular neuroscience provides the means by which people can *govern themselves*, in contrast to the need for top-down authorities that serve to command and direct. Health, wealth and happiness therefore become unquestioned social goals towards which we must individually work. As she remarks (2011, p 10):

> ... when individuals take up biological and psychiatric vocabularies for framing their own lives, these languages often dispose them to actively participate in various initiatives – ranging from the consumption of therapy, including medication, to parenting classes to productive workplace behaviours – all as part of their own pursuit of personal fulfilment and brain health.

For Thornton, the current era of brain culture has been driven by neoliberal social and economic formations. The focus on personalised responsibility and the retreat of the state are key features. Brain culture fills the void left by the state with an incitement to be independent, to choose mental health and to partake of a never-ending journey of recovery (Thornton, 2011, p 145). Sam Binkley (2011a) develops a similar account, building again on Rose's (1989) analysis of the role of the psy-disciplines (psychiatry, psychology, psychoanalysis, as well as disciplines of education and training) in the era of the welfare state. Rose's work explored how such knowledge and expertise was mobilised through the institutions of the family, school, factory, military and civil society in order to regulate subjectivity and to set norms that would be of value to the social good. Under a neoliberal political economy characterised by privatisation, deregulation and marketisation, however, psychological subjects have become 'pure resources in an environment of opportunity' (Binkley, 2011a, p 93). In this context, self-actualisation, self-management and self-maximisation become the maxims of psychological knowledge and the entrepreneurial practices that shape everyday life. Binkley has explored these aspects of brain culture through his analysis of the production of knowledge and expertise in positive psychology texts (Binkley, 2011b) and life coaching practice (Binkley, 2011a), noting the importance of examining the specific processes of subjectification within everyday life and popular culture.

Normalisation

The second process is *normalisation*, as exemplified by Thornton through an incitement for continuous self-optimisation found within popular accounts of neuroscience. Unlike 'normation', which, as Foucault described it, refers to the prescriptive delineation of normal behaviour, and 'trying to get people, movements and action to conform to this model' (Foucault, 2007, cited in Thornton, 2011, p 21), normalisation refers to the endless pursuit of an unachievable perfection. Rather than say 'you must be like *this*', normalisation invites you to 'be all you can be!' (Thornton, 2011, p 21). In any sphere of life, it would seem, popular neuroscience can help you to achieve an almost entirely interiorised self-mastery, enhancement, betterment. This is particularly evident in the fields of neuroeducation and workplace training, examined in Chapters Four and Five. This imperative for self-optimisation is bound up with what Thornton describes as 'healthism' and its counterpart, pathologisation. As such, brain health is 'a resource that can be acquired without limit', as demonstrated by a Nintendo computer game called *Brain Age* in which players can diagnose, calculate and improve their brain age (return to a younger self) by virtue of the game's carefully designed brain training exercise (Thornton, 2011, p 12). Here, health becomes a form of capital to be accumulated. As educational theorist Katherine Ecclestone (2012, p 464) has noted, the prevailing danger of this 'therapeutic ethos of culture and everyday life' is the production of a pathologically vulnerable subject – we are all potentially vulnerable and in need of emotional, social and psychological therapy (Ecclestone, 2012, p 475). Moreover, she argues that therapisation constitutes a new form of governance, using emotional pedagogies, positive psychologies and nudges to reduce social and political struggles to questions of character (Ecclestone, 2012, p 469). Where the desire for self-enhancement is interiorised, so, too, is the insurmountable responsibility for a failure to achieve it.

Cultivation of a neuromolecular gaze

The third process by which neuroscience shapes governance and citizenship is the *cultivation of a neuromolecular gaze*. In their detailed history of the neurosciences, Rose and Abi-Rached (2013, p 43) refer to the emergence of a 'neuromolecular style of thought'. They note that it may seem strange to begin their historical survey of the neurosciences as late as 1962, given the diverse investigations in neurology, brain disorders and the 'insane' mind that have preceded this date over the previous centuries. But their reasoning indicates a concern for understanding the broader significance of the neurosciences in shaping our behaviour, our identities, our philosophies and our very thinking about the neurosciences. They suggest (2013, pp 30-1) that:

> ... a new style of thought has taken shape, which has provided the possibility for the interdisciplinary neurosciences to achieve their current status, to move out of the laboratory and the clinic, to become a kind of expertise in understanding and intervening in human conduct in many different practices.

In summary, this style of thought holds that: the human brain is a biological organ which shares evolved characteristics with other animals; it is possible to reduce all neural process to molecular events (these processes being instances of chemical and electrical transmission); different parts of the brain have different evolutionary histories and functions, the activity of which can be visualised; and all mental processes correlate with specific and 'potentially observable material processes in the organic functioning of the neuromolecular processes in the brain' (Rose and Abi-Rached, 2013, p 43).

This necessarily reductionist methodology was required in order to map the brain, behaviour and 'normal' mental states. Investigations of complex social and cognitive occurrences and experiences at the level of basic molecular and algorithmic processes have flourished. In a sense, it is this very style of thought, reductionist, molecular, concerned with basic brain function, which hampers the potential applications of neuroscience in the 'real world'. The broad social and political implications of the basic functions of, for instance, the visual processing system, neurotransmission at the level of the cell, or the detailed chemical or electrical interchange of synaptic communication, are far from self-evident. But for Rose and Abi-Rached, this did not stop the neuromolecular style of thought from seeping into our everyday culture through specific forms of behavioural governance. This was particularly the case within psychiatry and the pharmaceutical industry, where the equation of mind and brain has opened the way (and lucrative markets) for the chemical treatment of mental disorders, and more recently the chemical enhancement of cognitive processes. But what really unlocked the door between the lab and life was a later set of developments relating to neural plasticity, which has taken hold as recently as the 1990s, and has been popularised by best-selling books such as *The brain that changes itself. Stories of personal triumph from the frontiers of brain science* (Doidge, 2007). Posing an apparent challenge to the idea that the brain is 'hard-wired' to do anything (although popular neuroscience enthusiasts see no contradiction in referring to the brain as both plastic *and* hard-wired), the notion that the brain changes biophysically throughout life, is shaped by environmental and social influences, events, upbringing, nutrition, even culture, paves the way for any number of therapeutic and policy interventions. If stroke victims can be rehabilitated, then cannot citizens be rehabilitated, or re-invented too? It is by virtue of this neuromolecular style of thought, supplemented by the promise of plasticity, that 'brain training' technologies and products have been touted, new programmes for governing have been developed, and new spatial design practices for shaping behaviour have evolved. In many cases these practices draw on contested neuroscientific insights, as the next chapter considers. We witness here not the 'breaking out' of new neuroscientific realities into popular cultural texts, but more the co-evolution of accounts of the self that fit into existing interpretive schemas and that are significantly buoyed by the rhetorical power of the neuromolecular gaze.

Governing through the brain

By cultivating a neuromolecular gaze in which human experience is reduced to basic algorithmic brain processes, and by simultaneously correcting this biological determinism with the optimistic hope of brain plasticity, a space is opened up through which each and all can govern themselves, self-optimise, and be governed *through* their brains. This fourth process is evident in the moral lessons we are to take from various aspects of popular cultural reference, whether in brain training video games, media reports of new brain scanning discoveries, self-help literatures, parenting manuals or life coaching courses. While such texts may all have subtle, long-lasting and cumulative effects on the narratives we tell about ourselves, the norms we set and the goals we desire, the more immediate process by which political actors might govern through the brain should also concern us here. The rise of the behaviour change agenda in UK public policy has been outlined elsewhere (see Jones et al, 2013, for an extended discussion). This agenda is based on the perceived need to condition citizens' choices in order to overcome our inherent cognitive biases, which include our susceptibilities to psychological priming, our short-termism, and our vulnerability to act on emotional impulse. There are debates, too, as previously noted, about how best to cultivate 'good citizenship' by building on our innate sense of empathy, collectivity and altruism (Taylor, 2007), by liberating us from our ingrained habitual thinking through more self-possessing mindfulness techniques (Rowson, 2011), or by shaping our character through forms of educational and managerial practice based on positive psychology. In this regard, governing through the brain requires a certain amount of governing the self, in the manner outlined by Thornton. But this doesn't just happen in isolation; it requires infrastructure, investment, sometimes new legislation *and* the legitimising force of objective scientific authority. Hence it is not only through working on the self that governing through the brain is realised – it entails intervention; it is the governing of others enabled by both the popular common sense of brain culture and the scientific expertise of those who know and understand the brain.

Again, this has far-reaching implications for governance and citizenship. For example, 'early intervention' programmes, which focus welfare policies on the first three years of children's lives as critical developmental windows through which future behaviour is said to be shaped, are gaining ever more political traction in the US and UK. Rose and Abi-Rached (2013, p 196) argue that despite all their scientific groundings, such programmes do not identify children at risk of developing behavioural problems by biological, neurological or genetic signs. There is no widespread screening of behavioural disorders (although controversially, fMRI scans 'for ADHD' [attention deficit hyperactivity disorder] are now advertised in the US). Rather, children whose families have come to the attention of social services are furnished with programmes to target and reduce such risks. These programmes are not neurological, but social – involving intensive training for parents in dealing with behavioural misconduct, improving reading and readiness for school.

There are no new brain-based *solutions* here, although neurological explanations abound. Rather, such governmental programmes are part of a continuing history of attempts to *shape the good citizen* through managing family relationships, targeting so-called 'cycles of deprivation', and *thereby* solving social ills – leaving structural inequalities intact. Rose and Abi-Rached do not dwell at any length on how such processes might re-invent citizenship, possibly because the impacts of governing through the brain are in their relative infancy. But they do draw our attention to the likely impact of brain culture itself on policy-makers, suggesting that there is a 'growing belief, among policymakers and in public culture, that human neurobiology sets the conditions for the lives of humans in societies and shapes human actions in all manner of ways not amenable to consciousness' (Rose and Abi-Rached, 2013, p 226). Notwithstanding the endurance of psychological accounts of the self and our refusal to be subjected to a neurobiological determinism beyond our conscious perception, this growing belief also sets up the conditions in which social problems and social ills are re-imagined as 'bio-social', rooted in the molecular biology, chemistry and physiology of the brain, and solvable only by the corresponding production of the bio-social subject and by the subjectification of the figure of the 'neurocitizen'.

Subjectification

Rose and Abi-Rached briefly suggest that the neurosciences herald a new 'mode of subjectivation [subjectification]' (2013, p 233). But far from suggesting that the neurosciences *produce* subjects, posing a radical threat to personhood, they draw attention to the limitations, continuities and paradoxes inherent in this complex relationship between brains and people. They recount how neuroscience emphasises non–cognitive processes, automatic behaviours and habitual tendencies. While it may appear to challenge our view of ourselves as autonomous decision-making subjects, this position is seen to sit quite happily alongside 'ideas about choice, responsibility and consciousness that are so crucial to contemporary advanced liberal societies' (Rose and Abi-Rached, 2013, p 21). As such, the neurosciences are no threat to the idea of a willing self, but rather incite citizens to pre-empt their own potential shortcomings, to work with and compensate for their frailties, to manage and improve their brains. At the same time, through the neuroscientifically informed techniques of population management, governments will render the brain calculable, bring to light our perpetual vulnerabilities, risks and susceptibilities, and intervene to control for our future actions and future 'disorders' of the brain. In this regard, Rose and Abi-Rached (2013, p 221) do not so much fear the coming of 'brainhood'. Rather, they see the legacy of at least a century of *psychological* rather than neurological conceptions of the self at work in popular culture and everyday experience. It seems that despite the apparently undisputable validity and expertise of neuroscientific accounts of our habitual, irrational, non-cognitive selves, and regardless of the best efforts of critics of the great neurological revolution, we stubbornly refuse to be determined by our brains. At the same time, it seems that we are all too willing to accept a deeply psychologised account of our individual selves and responsibilities.

The specific mechanisms of subjectification are complex and hotly debated, and it is these precise processes that are the core focus of this book. It goes further than an exploration of the production of knowledges of personhood in academic and popular texts in order to more fully examine actual people in specific places. Its politically engaged account specifies how people become citizens through the particular governing modalities of brain culture in context, through a political analysis of the brain world itself. Engin Isin's (2004) original account of contemporary citizen subjectivity is instructive here. Isin argues that the modern autonomous, rational and choosing subject of liberalism and neoliberalism has been superseded by a neurotic citizen. Unlike her (neo-)liberal predecessor (the citizen of Thornton's notion of brain culture), the *neurotic* citizen is not expected to internalise the social goals of health, wealth and happiness. She is not obliged to calculate and calibrate her behaviour to these ends through self-discipline. She is no longer governed through her freedom, through the active technologies of the self to which she subjects herself. Rather, it is within the context of what Isin denotes (after Deleuze) as contemporary control societies – characterised by the increasing securitisation of the state, surveillance, border control, exclusionary practices, cultures of fear, divisions between 'us and them'– that the neurotic citizen emerges as a subject who must render her behaviour according to anxieties and insecurities rather than rational choices (Isin, 2004, p 223). She must manage her anxieties, for they will not be cured.

While the (neo-)liberal citizen emerged as subject to the disciplinary effects of the scientific management of economy, population and society, the neurotic citizen is the product of a medicalised society, 'an age of neurological explanations with genetic structures, neurochemical causes of "fear acquisition", neurocircuitry, neurochemistry and neuroendocrinology of anxiety disorders' (Isin, 2004, pp 225-6). Thus Isin coins the term 'neuroliberalism' to describe the rationale for tranquilising emotionally anxious citizens within this context of insecurity, fear, threat and imperfection. These contextual particularities are part and parcel of a political-economic situation in which citizens are no longer held to be rational people. The highly medicalised, neurobiological account of citizen irrationality critiqued by Isin can be identified as one of the characteristic features of the programmes of policy and governance associated with the brain world, and his critique of neuroliberalism informs the standpoint of the book.

About *Brain culture*

The central concern of this book is to explore a range of contemporary spheres of policy and practice in which neuroscientific, behavioural and psychological research is forging new directions. In grouping together these three sometimes distinct, sometimes overlapping, forms of knowledge, the aim is to investigate the political ramifications of actually existing brain culture. Brain culture refers not only to the contested *impacts* of these research areas for our philosophies of mind and our sense of human identity, nor only to the *application* of research insights for which the 'neuro' prefix has come to signify. Rather, brain culture describes the *circulation*, within specific socio-spatial contexts, of a particular approach to human identity, sociality, decision-making, willpower, reasoning and responsibility, which is shaping how citizens govern themselves and how they are governed by others. The book is not solely concerned with whether neuroscience, behavioural science or psychology are legitimate grounds for governing citizens, specific social groups or whole populations. Instead, it is about the lived experience of citizens in the brain world, examined through three spheres in which brain culture is evident in the UK: neuroarchitecture, neuroeducation and positive psychology in the workplace.

Some further notes are necessary on the scope of the book. In foregrounding brain culture as something which is already 'out there' in several spheres of policy-making and practice, the book makes only passing reference to the insights and research programmes of contemporary and earlier neuroscientists on which the notion of brain culture has been built. Several of the titles reviewed in this introduction do the work of describing the precise histories of neuroscientific lab experiments, research programmes, disciplinary developments and recent discoveries, and this work is not repeated here. The reader may also be concerned that the neurosciences and behavioural sciences, including psychological strands of economics, are being too easily conflated in the narrative provided here. There are important distinctions between the mechanistic models of neuroscience (the search for the neural drivers of behaviour) and the insights of behavioural economics, which largely aim to describe the divergence of observable behaviours from a norm. And that is not to mention the highly diverse field of psychology that encompasses a broad spectrum of competing models of the mind. There are significant differences between each of these research approaches, in their theories and methods alike. But their shared endeavour is to seek explanation for human experience in the decision-making capacities of the brain.

The book does not investigate the related but specific influence of psychoanalytical theory on public policy, governance and citizenship. For analysis of the contested role of psychoanalysis in governing subjectivity and citizenship, Miller and Rose (1988) and Frosh (2001) provide significant contributions. For research, policy and practice influenced by psychoanalytical thinking, human decision-making is necessarily a less scientific affair (and its science status highly contentious). Its impact on contemporary public policy and practice is also arguably less explicit. While psychoanlaysis is an important background component of the psycho-spatial analysis adopted in this book, and while there are clearly parallels between a psychoanalytic attention to the irrational, personality, habit formation and the unconscious, it has not quite become the cadre of emerging knowledge used as evidence of the causes of human decision-making in the same way as have aspects of the behavioural, psychological and neurosciences which are the focus of this book.

In appealing to changing popular cultural understandings of narrowly conceived decision-making capacities, and the authority of the neuro- and behavioural sciences, it is no wonder that those advocating public policies based on the brain can claim to be politically neutral. Whether it is the mildly regulated free market of libertarian paternalism (aka

Nudge), or a social democracy built on neurological reflexivity (as promoted in the work of the RSA), it seems that brain culture may help us to discover the elusive electoral safe ground of the *real* Third Way (Grist, 2009). There are three crucial problems with this explanatory framework that inform the approach taken in this book. The first is a problem with scale. A focus on the neuromolecular, the brain, or behaviour necessarily provides a blinkered view of the determinants of human action. Attentions are often narrowly focused on detailed biochemical processes and neurophysiological functions. The second problem relates to context. The decision-making sciences may well be attentive to the genetic determinants of conscious thought, the responses of the brain to environmental stimuli, conditioning, even the influence of particular 'choice architectures' which guide our decisions (Thaler and Sunstein, 2008), but it is rare that neuroscience, behavioural science or psychological science sufficiently engages with the deeper contexts in which people exist and act. Neuroscientific methods do not lend themselves to an understanding of the person as situated in history and in geography, shaped by an over-determined myriad of economic, political, cultural, social, technological and environmental contexts. Third, the decision-making sciences do not have the means to account for their own role in the shaping of the people whom they seek to understand – for there is nothing natural about their (collectively produced, well-funded, institutionalised, technologically enabled) knowledge in this brain world. These problems of scale, context and situatedness are explored in detail in the following chapter, which also sets out a geographical framework for better situating the 'neurocitizen' in her psycho-spatial context.

One final problem is a more fundamental one that informs the rationale of the book, although it is so intractable as to be unwise to tackle it head on. It is the deceptively simple observation that scientific truth *changes*. One scientist's breakthrough in the world of phrenology or vibratiuncles, for instance, is soon the pseudo-science of yesteryear. While these may be crude examples, the point is that the scientific endeavour is a highly unstable one, subject to refinement, refutation and ongoing dispute. To appeal too heavily to scientific authority in political decision-making is therefore to overclaim the certainty with which scientific facts are held to be true at any one time. Many have argued that the popular media play a crucial role in perpetuating such overstatements with regards to neuroscience (O'Connor et al, 2012; Satel and Lilienfeld, 2013). One need only point out, as critics such as Zerilli (2013) and Tallis (2011, p 326) have done, the historical inhumanities of political scientism (from eugenics to racist craniometry) as a warning against a reliance on the human sciences as a blueprint for how we should live together.

Seymour and Vlaev (2012, p 450) have argued that there is little evidence of neuroscience in use in actual policy. Yet several other authors reviewed in this introduction have warned that we may be at a turning point in terms of the influence of brain culture on governance and citizenship. Behaviour change, early intervention (in family welfare), education, mental health, military training, marketing and the law have been marked out as areas in which brain culture has already taken hold. But clearer empirical evidence is needed if the claim that policy-makers have been influenced by neurobiology is to be upheld. *Brain culture* thus searches out instances where the circulation of brain culture can be said to be having an important impact on policy and practice, and to identify how this might be changing governance and citizenship in the UK. As such, the book is interested in the circulation of brain culture, its interpretation in practice, and its role in mediating experience. Exploring the terrain in between the scientific lab and the practical chalkface, it investigates the processes by which the 'neurocitizen' is constituted as both a subject of the brain world and as an object of neurologically informed governance in specific spheres of policy and practice. Each chapter thus revisits, to different degrees, the processes outlined in this introduction: self-governing, normalisation, the cultivation of a neuromolecular gaze, governing through the brain and subjectification.

The book brings together three distinct spheres (neuroarchitecture, neuroeducation and positive psychology in the workplace) in order to develop an account of the implications of brain culture for governance and citizenship. The following chapter offers a critical account of brain culture based on developments within human geography with the aim of establishing better understandings of the brain and behaviour *in context*. It outlines several geographical interventions on the environmental determinants of human behaviour, the whereabouts of human subjectivity, and the context of political agency that each offer contrasting viewpoints on the kinds of political critique that novel policies and practices characteristic of these three spheres of brain culture might be subjected to. It offers a challenge to the certainty by which neuroscientific, behavioural and psychological insights are used to justify particular political and policy agendas in these spheres. The book contends that the neurosciences have tended to over-claim how they might radically change our understandings of ourselves, our societies, and how we should live. In so doing, they risk neglecting seemingly more mundane and more geo-historically specific aspects of how our situations and contexts shape our minds and thoughts, human experience, subjectivity and political agency. It is important therefore not to over-claim the insights offered by a human geographical analysis of the brain world. The purpose instead is to examine the partial nature of those accounts offered by neuroscience and geography alike, and to put into question the apparently unequal status of these contrasting ways of knowing human behaviour and action. (One does not, after all, hear the research insights of human geographers aired in the media in quite the same way as new neuroscientific 'discoveries' are revealed, and there is no multi-billion euro funding for the Human Geography Project!)

The first field of brain culture considered in Chapter Three explores the knowledge and practices associated with the new field of neuroarchitecture. ANFA was formed in San Diego in 2003, and a handful of books, and many more articles, have been published on the subject since this time. This novel interdisciplinary approach integrates the established tradition of environmental psychology with approaches from artificial intelligence, systems computation and neurobiology. Neuroarchitecture seeks to establish applied approaches to urban planning and design that take into account both the effects of the built environment on brain activity, and the neurological correlates of space perception – the 'push' of the world around us. This chapter outlines the organising concepts and methodologies of the small field of neuroarchitecture, and places it in the much broader context of environmental psychology and urban design disciplines. Examples are given of policies and practices associated with designing spaces to improve learning and to shape health-related behaviours. The chapter examines how a particular spirit, ethos, moral and political stance is engendered through the disposition of people, things and attention in space, and identifies the democratic problems posed by a potentially *brain-based spatial rationality*.

Chapter Four investigates neuroeducation and brain-based teaching models. These practices have been subject to considerable public debate regarding the controversies of 'smart drugs' or fish oils used to improve exam performance, the psychopathologisation of children with ADHD, and the apparently enthusiastic adoption of commercial Brain Gym and cognitive training programmes by teachers and schools. A significant amount of educational research has embraced the turn to the cognitive learning theories and the learning brain, while UK government-funded research programmes have also attempted to sort the 'neuromyth' from scientific reality. The adoption of neuroeducational practice in schools is as yet unknown in scale and scope, but it is widely heralded in its future potential. There is an active field of conferences, networks, events, publications and consultancies that promote neuroscientific approaches to learning. This chapter demonstrates the political, cultural and institutional conditions that have made the adoption of neuroeducational approaches possible within contemporary schooling in the UK. It examines how educational psychologists are developing new roles as sceptical intermediaries between neuroeducational research and school-based practice. A brain culture in schools has unintended consequences for the processes by which children and young people are subjectified as learners, and as citizens whose behaviour is to be governed. It is argued that neuroeducation does not properly situate diverse learners in terms of the geographical contexts that make up the learning self. Neuroeducation thus remains blind to broader historical and cultural transitions as well as the specific neighbourhood geographies of unequal schooling that produce specific norms for learners, citizens and subjects. In so doing it risks forwarding problematic medicalised explanations of learner behaviour and reductionist visions of how learning works.

Another sphere through which brain culture can be discerned is in the workplace. Here neuroscience and positive psychology have influenced the discourses and embodied practice of human resources (HR) management, and workplace training programmes. Chapter Five maps the expansion of positive psychology consultancies within the UK, and their relationship to specific developments in the academic 'movement' of positive psychology and the field of organisational studies in the US. The chapter outlines recent research advances in positive neuroscience and organisational neuroscience, and traces their influence on a growing world of organisational management, leadership and HR approaches, Buddhist-derived mindfulness practices, self-help techniques and workplace wellbeing strategies. Psychometric testing, performance management, strengths-based self-assessment and emotional management have become de rigeur in workplace settings, but rarely does analysis of these work-based practices appreciate the governmental logics associated with orchestrating positive affects. When viewed as part of a broader happiness agenda in public policy, the political and economic rationalities of such workplace training programmes becomes clearer. The chapter examines how the scientific credentials and business rationales of positive psychology in the workplace share a specific history. This co-evolution of knowledge and practice enables practitioners and consultants to provide evidence-based promises to do 'less with more' in a difficult economic climate. Workers are urged to cultivate self-mastery and optimism in strategic alignment with organisational goals in order to optimise themselves and their levels of productivity. In turn, the workplace becomes a site of entrepreneurial citizenship through which a very culturally specific set of norms, virtues and behaviours is promoted, with brain-based explanations apparently attesting to their timelessness, neutrality and universality.

By investigating policy, practice and research within these three worlds, the book provides a new account of the emerging influence of the neuroscientific, behavioural and psychological insights across distinct but interrelated social spheres. In this sense, the book draws out some of the broad implications of actually existing brain culture in everyday urban spaces, schools and workplaces for shaping our relationships with our brains, selves and each other. It identifies the specific impact of brain culture on aspects of citizenship and governance that have hitherto been under-explored. It is indebted to a growing field of critical research on the sociological, political and cultural influence of brain culture, and brings these critical theoretical debates to bear on a grounded analysis of contemporary policy and practice. The book demonstrates the value of taking a geographical approach to investigating this brain world by considering the explanatory potential of understanding the geographical *situation* of the citizen to be governed through brain culture, and the *context* in which brain culture is becoming established as a privileged account of human behaviour and as a rationale for action. It argues for an appreciation of the partial and geo-historically contingent explanatory frameworks offered by the neuro-, behavioural and psychological sciences in order to forward a more cautious approach to their application in policy and practice. In so doing it carves out a space within the emerging brain world for valuing both the knowledge and doubts expressed within the humanities, arts and social sciences towards explaining the ever-changing, diverse and situated human condition.

A note on methods

Chapters Three, Four and Five draw on original research interviews undertaken in 2013 with neuroarchitects, neuroscientists, neuroeducators, educational psychologists, positive psychology practitioners and management consultants based in the UK and US. A total of 50 interviews were conducted, although not all of these interviewees are cited in this text. The interviews were semi-structured qualitative interviews that lasted on average an hour each, and were conducted by the author and two research assistants. The interviews were transcribed and analysed using a coding framework that included a mixture of inductive and deductive themes. The interviews have been anonymised here, with the exception of Chapter Three, where permission has been given to name the interviewees. The interviewees have given their informed consent to participate in the research. Other sources of data on which the book is based include policy documentation relating to education, wellbeing and welfare policies, peer-reviewed journal articles and academic texts, textbooks, professional publications and the websites of various research networks, institutions, not-for profit organisations and commercial consultancies.

TWO

Brain culture in context

A psycho-spatial analysis of brain culture: understanding context

How do neuroscientific, behavioural and psychological accounts of everyday behaviour reconfigure our sense of citizenship? Responding to this question means taking seriously the historical and geographical *situation* of the citizen who is governed through the brain, and the social and political *context* in which brain culture has emerged as a way of explaining the human condition. This chapter sets up the analytical framework pursued in the remainder of the book. It describes how various spatial insights from the discipline of human geography present an original perspective from which to politically analyse policies and practices influenced by brain culture. It is argued that the subsidiary endeavours of behavioural geography and environmental psychology, non-representational geographies and Foucauldian geographies of political rationality generate new questions regarding the role of brain culture in contemporary practices of governance and citizenship.

Principally, the chapter investigates the *whereabouts* of subjectivity, and argues for a geographical approach to analysing those practices and policies which have been influenced by brain culture. It extends the approaches of those pursuing psycho-social analysis who refuse to separate the subjective and psychological from social context. The two are seen as mutually constitutive within the psycho-social approach. Some of these approaches integrate psychoanalytic understandings of the relationship between the conscious and unconscious, with an attention to the material, social and cultural structures that shape and are shaped by human subjectivity and action (see, for example, Walkerdine et al, 2001). Others are concerned with understanding precisely *how* everyday social practices are implicated in subject formation and the production of social relationships without recourse to the supposedly hidden psychic drives of the unconscious (see, for example, Wetherell, 2012, p 134). It is this latter focus on the 'organisation and normative logics of the unfolding situated episode, context, interaction, relation and practice' – with an additional emphasis squarely on *where* everyday social practices co-constitute subjectivities (particularly citizen subjectivities) – which informs the psycho-spatial approach developed in this chapter.

I begin by briefly exploring how the spatial tropes of situatedness, context and environments are addressed in existing accounts provided in social neuroscience, popular psychology and their critics, before introducing the contribution of human geography to these debates.

The main argument forwarded here is that there is a need to take seriously a geographical approach to 'situated subjectivity', and that such a sensitivity has implications for how we understand political agency and citizenship in the context of brain culture. Fears of political manipulation through the neurosciences and behavioural sciences have been widely expressed as much as their potential for political empowerment has been heralded. Yet these fears and hopes are reliant on a rather simplistic conception of the relationship between mind and environment. In this chapter, I therefore elaborate on the contribution of geographical thought to better understanding the crucial importance of a widely conceived and scaled-up notion of context on decision-making and of the political rationality of the brain world itself on shaping subjectivity in a circular manner. Human geography offers a range of theoretical positions from which to investigate brain culture, many of which have themselves taken a 'neural turn'. The chapter argues that those approaches which can bring together an emphasis on *both* the vital materiality of human life and its discursively constructed nature offer the deepest insights into the realms of brain culture investigated in the remainder of the book.

Many have argued that the behavioural sciences and related disciplines are being used to promote morally questionable forms of intervention in the governance of citizens (Furedi, 2011; Saint-Paul, 2011). In contrast, the RSA, an organisation noted in the previous chapter for its embrace of the neuroscientific 'enlightenment', argues that brain culture will be politically empowering, that knowledge about the brain will help people to develop self-knowledge and self-mastery (Rowson, 2011). Both these divergent positions suggest that the citizen will internalise social, political and therapeutic practice, that the social 'outside' somehow gets into the person. Within neuroscience there is increasing interest in the influence of cultural difference on the brain (see, for example, Chiao, 2009). Social approaches to neuroscience have also been growing since at least the 1990s (see, for example, Cacioppo and Berntson, 1992), with a recent emphasis on understanding pro-social behaviour (see, for example, Crockett, 2009). Understanding the impact of 'environment' and gene–environment interactions on the brain has also been a long-standing concern, for instance, in research that has examined the interaction between genes and upbringing in shaping mental disorder (see, for example, Caspi and Moffitt, 2006). Epigenetics, which examines the modification of gene expression through the life course in response to environmental context and upbringing (particularly in the maternal relationship), is a booming area of research. As Rose and Abi-Rached note (2013, p 50), these research priorities are complemented by theories of brain plasticity, by which the brain changes, neural network connections are generated or threatened, and even new neurons are produced. This is suggestive of the external environment getting inside the very materiality of the brain.

But, as Abi-Rached and Rose (2013) point out, the notion of the 'environment' is as ill defined in political sociology as it is in neuroscience, meaning that debate between the two relates to two entirely different scales of explanation and is therefore antagonistic. While a neuroscientist might posit, for instance, that the problem of anti-social behaviour is caused by the impact of early neglect on the brain, for the majority of sociologists, social dysfunction is best understood at the level of social structure, of growing up in a context of poverty, inequality and injustice. There is a consequent gap in current conceptions of just what 'context' means in relation to the brain and behaviour, and how we might better understand 'brains *in situ*' (Abi-Rached and Rose, 2013, p 227). The project of developing a critical neuroscience has embarked on addressing this issue by focusing on embodied experience and situated subjectivity. Choudhury and Slaby (2012, p 11, emphasis added) contend that 'mental processes' should be 'understood as *constitutively* embodied and environmentally embedded such that they cannot be properly characterised without reference to their bodily dimensions and relations to the physical and social environment.' For them, any separation between perception and action, any notion that the brain simply cognises environmental inputs and then processes them as behavioural outputs, is false. Instead, like the philosophers of mind encountered in Chapter One (Susan Hurley, Alva Noë and Raymond Tallis alike), they propose that human experience is characterised by 'enaction' – actions are as much shaped by the world as they are constitutive of it. To an extent, this position is reflected in Tallis's (2011, p 350) contention that we should approach the brain as 'inseparable from a community of minds and the worlds that its component selves have built.' Albeit Choudhury and Slaby are putting to use a set of theoretical insights from post-structural theory that Tallis derides as equally destructive of human subjectivity. Yet both agree that conceptualising the body's role in consciousness is a crucial endeavour. One of Tallis's most well-known contributions to philosophy has been to argue that what makes us uniquely human is not that we have more developed brains, but that we use our *hands* as tools through which to directly engage with and manipulate environments. Such a focus on situated and embodied consciousness and experience is emerging within some contemporary models of neuroscience. Choudhury and Slaby (2012, p 10, original emphasis) identify the "'4EA approach": the mind as *embodied, embedded, enacted, extended, and affective* (after Protevi, 2009).' And Tallis (2011, p 352) describes the 'DEEDS approach', which sees cognition as 'dynamical, embodied, extended, distributed and situated' (after Wheeler, 2005). Their point is to understand brains *in situ*.

These recent developments resonate with the observations of feminist biologist Hilary Rose (2004, p 59), who similarly identifies shortcomings in how neuroscientists tend to objectify consciousness entirely out of context. In her discussion of the limits of neurobiology, Rose maintains that this recent interest in consciousness erroneously locates it within the individual organ of the brain. Neuroscience thus ignores the established contributions of the humanities and social sciences to understanding consciousness as 'subjective and intersubjective within a historical context'. 'For social theory', she argues 'there can be no development of individual consciousness without a social context.' Hence social movements such as the black, class and feminist consciousness movements emerged not because of, one might speculate, aggregated improvements in the protagonists' improved perceptual capacities to discern injustice, their synaptic plasticity to change ingrained habits of thought, or their improved neural motivation to take action. Of course, activists and publics use their brains, but these social movements were indelibly shaped by social practices of meeting, discussion, awareness-raising and committed action in socio-political contexts of exploitation and oppression (Rose, 2004, p 63). In what is now becoming a familiar accusation of contemporary neuroscientists' likeness to phrenology, Hilary Rose outlines how prominent scientists (Susan Greenfield and Steven Pinker among them) have ignored context in their attempts to universalise consciousness. They have equated consciousness with cognition, or have regarded it as simply the opposite of unconscious/asleep or indicating mere 'awareness'. Their attempts to locate complex social phenomena in specific brain regions are thus misguided. This 'reductionist materialism', argues Rose (2004, p 68), 'simply eschews context' and overlooks the cultural specificity of our conceptions of social phenomena in favour of biological universals. By contrast, it is possible to take a non-reductive perspective which values the role of the embodiment of consciousness without recourse to a naïve sense of materialism/physicalism. When the brain as body is considered simultaneously as a fleshy object and a discursive construct, contextually specific social and political relations are brought into view. Naturalised and universalised truths based in science are questioned, and new uncertainties and hesitations are brought to bear on how we should live and what we should do.

This is not to say that neuroscience and psychological science *tout court* can be characterised as straightforwardly reductionist, as might be supposed. Recent research on schizophrenia, for instance, has reported that the condition is highly context-dependent, such that it has been commonplace to explore why there may be higher incidences in urban areas (Pedersen and Mortensen, 2001; Krabbendam, 2005), and among migrant populations (especially, it is noted, from countries where the majority population is black). While the causal relationship remains unknown, hypotheses range from an increased risk of exposure to early childhood viruses in urban areas, to stresses associated with social deprivation, racial discrimination and poor housing. According to one study (Selten and Cantor-Graae, 2007, p 9), the risk of developing schizophrenia may also be connected to 'the chronic experience of social defeat', experiences of subordination and exclusion which may affect the dopamine system in the brain, a system often associated with reward, addiction, motivation and behavioural modification. This signifies the emergence of an important neuroscience research programme which seeks to integrate contextual and environmental factors, particularly 'urbanicity' and migration, with neurochemical processes. That the researchers recognised the conceptual limits of their findings, stating that '[t]he experience of defeat is in the eye of the beholder' (Selten and Cantor-Graae, 2007, p 10), suggests that there are considerable methodological and epistemological difficulties in measuring and accounting for such subjective contextual phenomena. It is not clear, for instance, how well the lab studies they draw on to make the connection between social defeat and the dopamine system can really be said to say anything about the actual human experience of social defeat, relating as they do to seeing how much cocaine monkeys consumed when housed individually or socially, and introducing rodents to more aggressive male counterparts. A certain reductionism remains in the comparison of urban life with the reactions of incarcerated rodents.

A radically different explanation of schizophrenia and other mental disorders offered by sociologist Alain Ehrenberg speaks of a contrasting conception of context. He has argued that even apparently universal and biomedical diagnoses of schizophrenia and autism *become* diseases in the context of burdensome societal expectations of autonomy: 'a condensed and radical expression of the problems of socialisation and sociality experienced by an individual in a society founded on autonomy' (2011, p 135). In the contemporary economy, service work is characterised by a permanent pressure to interact with and attend to customers – risk-taking and initiative are valued above obedience. As such, he argues that it is not only mental health patients but every citizen who is compelled to be an entrepreneur of their own life. Failure to do so is acutely personalised to the neuromolecular and genetic scales, while mental healthcare shifts towards cultivating an historically contingent responsibility to achieve the skills and competencies to function autonomously as an individual.

From this sociologically contextualised perspective, we are a far cry from some of the ideas emerging in popular psychological accounts of context, and indeed, in behavioural science informed policy formulations of 'choice architecture'. These forms of policy shape the environmental contexts in which we make decisions in such a way as to make certain choices easier, more likely, more desirable – with multiple 'choice architects' setting out decision-making pathways in order to help us overcome our behavioural incompetence, psychological heuristics and self-defeating habits. Take, for instance, psychologist Sam Sommers' (2011) *Situations matter: Understanding how context transforms your world*. In this book, readers are promised a 'science of situations' (Sommers, 2011, p 9). '[O]rdinary contexts', Sommers writes, 'where you are, who you're with, what you see around you – transform how you act and, indeed, what kind of person you appear to be' (2011, p 8). But his notion of situation or context is one which is highly specific – limited to the immediate setting, both in terms of time and space, of a particular behaviour or interaction. This context ranges somewhat inexplicably from the incomparable contexts of having an argument in an airline queue, to the unsuccessful interrogation of prisoners at Abu Ghraib prison in Iraq. In the latter case we are told that it was only when interrogators started to shift their tactics 'from brutality to brains' (Sommers, 2011, p 10), by building rapport with the prisoners, that any useful military intelligence was gathered. Sommers outlines how situations affect how we interact, how we behave, how we might be too busy, too tired, too saturated with information to make the most appropriate decisions. But there is no sense here of *where* these decisions come from – what are the reasons given by people for making particular decisions, what are the rationalities behind decisions and actions? We are offered no explanation of how we culturally, socially and politically come to see ourselves in a particular light or act in a particular manner. Instead we are provided with a set of self-help lessons for appreciating and eliminating our susceptibility to context. Sommers therefore promises that 'this book will alter the way you think about human nature, thereby making you a more *effective* person' (2011, p 8, original emphasis), its lessons being aimed at improving both your professional and personal life.

Context clearly means different things in different academic disciplines, and the discipline of human geography is well-placed to challenge such partial conceptions of why situations matter. The remainder of this chapter therefore takes a broad brush tour of three areas of human geographic thought, which each have something significant to teach us about bringing together the social and spatial aspects of context. The dynamics of *human–environment interactions* (informed by behavioural geography and environmental psychology), *affective architectures* (as conceived by non-representational theory) *and contextual rationalities* (the focus of Foucauldian approaches) are elaborated as analytical starting points for a re-politicised account of neurocitizenship and governing through brain culture. While several disciplines are said to have had a spatial turn, human geography has ventured a somewhat lengthy and arduous route to its contemporary theorisations of space and spatiality. Tensions within the discipline between spatial scientists and those concerned with the spatial qualities of social relations remain largely unresolved.

It is not my intention here to rehearse these arguments. Instead, I outline three approaches to space within human geography which offer the most relevant insights for a context-sensitive account of the politics and practices of governing through brain culture. The aim is to sketch out some starting points from which to explore the as yet under-examined geographics of brain culture. As will become evident from the following narrative, the three approaches offered are as partial explanations of human behaviour as are the neuroscientific approaches already explored to this point. But their shortcomings provide valuable lessons for developing a spatial analysis fit for investigating brain culture in policy and practice.

Human geography provides a useful set of tools for understanding the specific situations shaping and shaped by human activity, the embeddedness of the mind in contexts, and the political context of brain culture itself. Yet the discipline is itself (like many academic endeavours, social policies and practices alike) subject to the influence of contemporary brain culture. During the 1960s and 1970s, behavioural approaches to geography enjoyed some popularity and became an integral part of the discipline's history. Influenced by psychology and a keen interest in perception and cognitive processes, behavioural geographers were concerned specifically with human–environment interaction – a focus which has been developed significantly by the contemporary field of environmental psychology. In some respects, these developments are characteristic of a humanities discipline that has long been in search of a science. Since the late 1990s, human geography has seen something of a discernable 'neural turn' through the development of 'non-representational theory' which has built on neuroscientific explanations of human action at the same time as geography has become more sensitive to materiality and the body in general.

Meanwhile, Foucauldian approaches have had a sustained influence on the geographical critiques relating to the truth claims of various forms of (often psychological) knowledge, techniques of governing and ethical practices of self-making or subject-formation. In the following sections, I examine geography's early encounters with behavioural psychology, and consider how this legacy has been sustained through the work of environmental psychologists. Second, I show how the non-representational approach in geography has been shaped by neuroscience, and outline some of its key stated political interventions. I devote considerable attention to this approach since it is a sub-field of geography that has demonstrated the most sustained engagement with neuroscience to date. I also consider some of the powerful critiques of this work offered by geographers and others, including challenges to the neuroscientific findings on which non-representational theory has drawn heavily in making strong claims about human action, willing and selfhood. I then proceed to explore the potential of geographical approaches to contextual rationality which stem from the Foucauldian theoretical tradition. I argue that this approach addresses some of the shortcomings identified in both behavioural geography and the non-representational project. Finally, I draw these threads together in order to offer some potential insights into a geographically informed analysis of brain culture.

Old behavioural geography and new environmental psychology

It may seem unusual to offer behavioural geography as a source of insights for discerning cutting-edge developments in contemporary neuroscience, since geographers' short-lived engagements with behavioural psychology between the 1960s and 1980s now occupy at best a marginal position in the discipline's history. Nevertheless, this transient foray has had lasting impacts in terms of highlighting to geographers some important conceptual missteps to be avoided, namely the spatial and political naivete of the behavioural geographers. Furthermore, the behavioural approach more generally lives on within environmental psychology, an increasingly influential field of study that informs public policy debates concerning the interrelations between the environment, mind and behaviour. This approach is particularly evident within urban design and the nascent field of neuroarchitecture explored in the next chapter. Behavioural geography's core concern was with the nexus between perception, cognition and action within specific environments. Scholars such as Julian Wolpert (1964), Bunting and Guelke (1979) and Golledge and Stimson (1987) were primarily interested in understanding spatial perception, exploring the relationship between images, revealed perceptions and behaviour, and modelling the way in which minds process information. In part a reaction to what they identified as the limits of positivist approaches to spatial science, behavioural geographers advanced geography in two main respects. First, they regarded space as more than simply a surface on which human behaviours were mechanistically played out. Second, they problematised the idea of a universal 'rational economic man' or *homo economicus*, considering instead actual behaviours at a disaggregated individual level in specific and constrained behavioural environments. This challenge to *homo economicus* has been notably central to contemporary (behavioural economic) explanations of citizen decision-making adopted within behaviour change policies.

The rapid demise of behavioural geography was precipitated by new developments in both Marxist geography, which saw geographers such as Doreen Massey (1975) and David Harvey (1989), as well as previous adherents to the behavioural approach such as Kevin Cox (1981, p 256), dismiss behavioural geography as 'just one more instance of bourgeois thought'. In these terms, behavioural geography was too empiricist, too descriptive and offered no real explanation for the deeper (structural) spatial contexts of human behaviour. At the same time, humanistic geographers (for example, Lowenthal, 1961; Buttimer, 1971; Tuan, 1971; Ley, 1981) took issue with the reductionist tendencies of behavioural geography, which seemed to give little time to the very human acts of interpretation, inscribed meaning, representations, actual experience, and conscious awareness of specific landscapes and environments. David Ley (1981, p 213), for example, argued that behavioural geographers were guilty of '[t]he absorption of man

[sic] into naturalist explanation [which] invariably implies reduction of what it is to be human.' In their cognitive models, any notion of meaning and interpretation, and the active role that humans play in constructing space and place, were derided as 'distorting effects' of the real science of spatial perception. From within behavioural geography, there were attempts to confront some of the criticisms relating to their de-humanising tendencies and blindness to structural contexts, and to develop the sub-discipline in new directions during the 1980s and 1990s. For example, Gold (1980, p 243) noted the limitations of behavioural geography: its concepts of cognition, perception and behaviour which were too loosely defined; its methods were not scientifically valid; it lacked relevance to issues of public policy; and it risked 'psychologism' – making inferences from individual behaviour and aggregating them into social patterns rather than examining the social phenomena themselves. This led Walmsley and Lewis (1993, p 15) to note that behavioural geography had simply replaced *homo economicus* with *homo psychologicus*, offering a severely limited analysis to environmental determinants of behaviour, and ignoring the fuller context of social, political and economic influences on conduct.

From a yet more critical, psychoanalytical perspective, Steve Pile, in recounting human geography's varied conceptions of the relationship between the body, space and the city, has probed the key texts in behavioural geography, examining how the field was 'constituted through an implicit, undisclosed, oppressive and obstructive masculinism' (1996, p 19). Behavioural geography, he argues (Pile, 1996, p 46), took the 'man' of 'man–environment' relations as given, failing to acknowledge the way in which the individual or subject is her/himself produced through discourse, is performed through situated practices, and is constituted through social relations. Such processes themselves remain 'unobservable' in the behavioural geographer's narrow categories of mental maps, attitudes and behaviour. In this sense, behavioural geographers ironically failed to develop an understanding of the mind itself, and as such succeeded only in reducing human experience to models of stimulus–response or information-processing. Pile (1996, p 36) therefore concludes that the subject produced by behavioural geography was an overly animalistic, computational and gendered 'ratomorphic' and 'robomorphic' man.[1] Behavioural geography therefore fell short in terms of its account of

---

[1]   This refers to the rat experiments conducted by Edward Tolman in 1948 examining the 'cognitive maps' of rodents which enabled them to build up, store and process information in order to navigate through their environments (Pile, 1996, p 36).

human behaviour because it offered no conception of the processes of subjectification. Nor was sufficient attention paid to the social, political, economic and cultural contexts in which personhood was discursively produced and through which people negotiated their identities in a world of social difference.

Despite the significant criticisms which have in many respects kept behavioural geography distinctly off the contemporary agenda in human geography, recent calls from economic and political geographers (Strauss, 2008; Jones et al, 2011a) for a more critical development of behavioural insights derived from human geography are worth consideration, not least because of the history that behavioural geography shares with the increasingly influential field of behavioural economics and the latter's role in shaping the behaviour change agenda in public policy, identified by Matthew Taylor (2011) as one key manifestation of brain culture. Indeed, there have been various attempts to 'revisit' behavioural geography (Cox and Golledge, 1981; Walmsley and Lewis, 1993), and to retrieve something of its potential explanatory power. It is attractive in its potential to predict spatial behaviour and to aid the design of new kinds of 'behaviourally savvy' urban environments. It also promises an understanding of human–environment relations within both 'natural' and 'cultural' landscapes, and in light of the changing global environment (human-induced climate change) and attitudes towards the environment itself. The recent work of Kitchen and Blades (2002, p 1), for instance, has argued for an integration of behavioural geography with environmental psychology in order to develop the techniques of cognitive mapping to explain how people acquire and process knowledge of their environment.

It is within environmental psychology that behavioural approaches have endured. While environmental psychology is firmly rooted in the methodologies and epistemologies of psychology rather than human geography, like behavioural geography, its purpose is to examine the interactions between people and the (natural or built) environment. The pioneering interventions of ecological psychologists Kurt Lewin, Roger Barker and Herbert Wright between the 1940s-1970s (including Barker and Wright's [1951] close observational studies of everyday behaviours in a school in Kansas, US) paved the way for an environmentally focused psychology. Lewin had argued for an understanding of the environment as a pivotal determinant of behaviour. Barker and Wright took up this challenge through their development of 'psychological-habit maps', which were intended to show how a person's field of possible actions were shaped by their environment (Wicker, 1979, p 3). Rather than focus on people's reported perceptions of environments, Barker and Wright were preoccupied by carefully documenting actual behaviours *in situ* – addressing what they saw as a major shortcoming in experimental lab-based approaches to psychology, which arguably still dominate psychological research today. This approach was radical in seeking environment-based solutions to problems that had hitherto been primarily seen as person-centred, such as mental illness or unemployment (Wicker, 1979, p 7). In observing behavioural patterns in particular behavioural settings, researchers aimed to identify the environmental *causes* of behaviours, and so pose solutions that tackled the 'larger environmental system' rather than apparently individual pathologies.

However, contemporary environmental psychology has long strayed from the approach of these early ecological psychologists. As such, environmental psychology has arguably been preoccupied only with negative human impacts on the environment rather than the active role that environmental settings play in shaping conduct, in a move that reverses the attentions of its founding scholars. The field is largely dominated by research that aims to contribute to environmental impact assessments, understand people's perceptions of environmental risk, and advise policy-makers on psychological strategies for encouraging pro-environmental forms of behaviour change. This has led some to observe that the discipline has become a 'psychology of sustainability' (Gifford, 2007 cited in Steg et al, 2012). The criticism here is that the relevance of this research to issues beyond those that relate more straightforwardly to the physical environment (for example, climate change, natural hazards, pollution, environmental management, the restorative and therapeutic benefits of 'nature') has been limited. The role it might otherwise play in conceptualising the co-constitutive relationship between *context* and human behaviour has accordingly been diluted.

Nevertheless it is worth dwelling for a moment on some recent debates and developments within the field of environmental psychology which may provide some useful pointers for the development of context-sensitive analyses of the policies and practices of brain culture. As Uzzell and Moser (2009, p 307) have noted, context should be of critical importance to the very venture of environmental psychology: 'there is no social environment that is not touched in some way by its physical context, and likewise every physical environment gains social meaning through culture.' In this sense, physical nature itself is always already mediated by social and cultural context, and it is a categorical error to assume that the mind and environment can ever be separate entities. Environmental psychology is arguably at a critical juncture, and Uzzell and Räthzel (2009, p 340) highlight a need to move away from a reductive focus on individual values, attitudes and behaviours in order to more fully appreciate the mutually constitutive role that social relations, and social contexts, play in shaping people. 'Individuals', they assert, 'are the sum of their social relations, that is, they are the cause and consequence of their relations to others and the environment.' Similarly, Devine-Wright and Clayton (2010, p 267) have argued that environmental psychology and notions of environmental identity must necessarily be understood by more concerted efforts to re-connect the self and the social:

> ... [b]ecause it implicates the self, identity has consequences for cognition, affect, and behaviour. Because it implicates the social, identity processes are embedded within wider, dynamic cultural, political and economic forces. Identity can be examined as both a dependent and an independent variable, both an effect and a cause.

Environmental psychologists themselves also recognise shortcomings in adequately defining what is actually meant by 'the environment', and are looking towards human geographers' accounts of socially produced 'nature' and the social production of space. According to Devine-Wright and Clayton (2010), this will develop a more 'transformational' purpose for environmental psychology which examines and shapes the more structural conditions in which people might live more sustainable lives.

The scientific methods and evidence-based approaches of environmental psychology have positioned such scholars as experts on sustainable consumption, environment and energy policies and effective strategies for behavioural change. However, the hitherto limited engagement of environmental psychologists with issues of social and spatial context, outlined by Devine-Wright and Clayton, poses particular limitations for attempts to analyse a brain culture that mobilises these very environmental psychological knowledges in order to shape citizens' behaviours. There is a political role played by environmental psychological evidence itself, in actively shaping brain culture, and in rendering citizens governable through their (*environmentally*) *psychologised* identities. Even at a basic level, ignorance of the social and spatial drivers of human action will lead to ineffective recommendations for behavioural change, which will be no guarantee of long-lasting cultural change. Such recommendations are likely to be disproportionately burdensome for those social groups that do not live within contexts in which their actions can be easily changed or new practices adopted. Furthermore, where the work of environmental psychologists is still preoccupied with naïve conceptions of 'natural surroundings' and space (in the limited sense of building design) as restorative/therapeutic/behaviour-shaping, their findings will also be limited in terms of scale. This certainly seems to be the case within the emerging practices of neuroarchitecture and behavioural design examined in the following chapter. In a sense, it is precisely the empiricist tendencies of much research in environmental psychology that puts limits on the scales of explanation that it can offer. In seeking to render measurable psychological 'traits', attitudes and values, and in its ambitions to operationalise workable models of behavioural change, what gets left out of many such studies is sustained consideration of the specific mechanisms through which human subjectivity is itself socially and spatially produced in specific contexts.

Non-representational geographies and a neuropsychological encounter with free will

Developed largely by geographer Nigel Thrift (2004, 2007), and continued by a group of his graduate students and many others since, non-representational theory draws on post-structural continental philosophies in the promotion of an approach to geography that focuses on embodied practices and the 'affective' (as opposed to cognitive) registers of human action. Non-representational geographers are concerned with those practices, events and relations that they argue are best understood outside of systems of representation (for example, structures of writing, language, conscious expression, interpretation and social constructs). In this sense, their approach sits well with a behavioural approach to geography which is interested in what people do more than what they think. It also promotes a vitalist conception of human subjectivity, emphasising the biological materiality of personhood. It sets out to problematise Western conceptions of personhood and its mistaken separation between biology and culture – rather, it is proposed that the individual, atomistic, rational self is a modernist fiction, and that we need to develop a new style of thinking concerning human nature. This new style of thinking has several effects: (1) it attends to those pre-cognitive/non-cognitive determinants of action that have been hitherto obscured by Cartesian dualisms between subjects and objects (specifically between minds and brains); (2) it develops an account of distributed subjectivity, in which the person is seen as a porous 'population of actors', a materially related coming together of human and non-human things at specific times (Thrift, 2008, p 85); and thus (3) it focuses on the processes and routinised practices by which the subject emerges in the performative (self-inventing and reproducing) encounter between bodies and things (Thrift, 2004, p 62; 2007, p 8).

It is through this avowedly post-rational approach that non-representational theory has gained a reputation for proposing a radically new 'spatial politics of affect'. One of the founding tenets of non-representational theory is based in contemporary neuroscientific accounts of personhood – that human action *precedes* cognition. Put simply, we act *before* we think. Thrift thus identifies a certain frailty in decision-making, which he derives from pioneering 19th-century psychologist Wilhelm Wundt, Benjamin Libet's neuropsychological experiments conducted since the 1960s, and subsequent narratives provided by neuroscientist Antonio Damasio and contemporary philosopher John Gray:

> Wundt was able to show that consciousness takes time to construct; we are "late for consciousness" (Damasio 1999: 127). That insight was subsequently formalized in the 1960s by Libet using the new body recording technologies. He was able to show decisively that an action is set in motion before we decide to perform it: the "average readiness potential" is about 0.8 seconds, although cases as long as 1.5 seconds have been recorded. In other words "consciousness takes a relatively long time to build, and any experience of it being instantaneous must be a backdated illusion" (McCrone 1999: 131). Or, as Gray (2002: 66) puts it more skeletally; "the brain makes us ready for action, then we have the experience of acting". (Thrift, 2004, p 67)

In Libet's experiments, research participants were asked to flex their wrists and to note the time when they felt the intention to move. This was then compared with readings from an electroencephalogram (EEG) which recorded brain activity in the cerebral cortex, signifying the person's readiness to move (the average readiness potential). This brain activity was consistently recorded at least half a second before muscle movement, and crucially, one-third of a second before participants reported their intention to move (Tallis, 2011, p 54), leading Libet to the controversial conclusion that the brain makes decisions before our conscious awareness of them, and thus that free will is a myth. Later research by Chung Siong Soon and colleagues using fMRI scanning to likewise investigate the 'unconscious determinants of free decisions' (Soon et al, 2008) has extended this one-third of a second delay between brain activity and conscious awareness of movement to five seconds (in the cerebral cortex), seven seconds (activity in the frontal cortex), or up to ten seconds if the time lag in fMRI recording equipment was taken into account (Tallis, 2011, p 55). Thrift takes up Libet's findings in asserting that consciousness control is an illusion (Thrift, 2008, p 88), and further, that this 'constantly moving pre-conscious frontier ... is highly political', a sphere of *microbiopolitics* (Thrift, 2004, p 67). Because it has become visible to science, this moment can be targeted and operated on by various 'entities and institutions' in order to produce particular political responses. This is particularly the case in urban design, which Thrift identifies as a new political field in which affective responses can be effectively pre-programmed. He offers 'design, lighting, event management, logistics, music, performance', 'cues that are able to be worked with in the shape of the profusion of images and other signs, the wide spectrum of available technologies, and the more general archive of events' (Thrift, 2004, p 68) as the broad means by which *affective architectures* can be engineered.

It is not easy to discern, let alone describe, just what the implications of a 'spatial politics of affect' might be for issues of citizenship and governance, although some sympathetic (B. Anderson, 2012) and several critical interventions have helpfully gestured at what might be politically at stake (Thien, 2005; Tolia-Kelly, 2006; Barnett, 2008; Pile, 2010). Nevertheless, in order to try to understand what a geographical contribution to understanding brain culture might look like, we must necessarily examine avenues of enquiry which have already been proposed for integrating a neuroscientific and geographical approach. Non-representational theory continues, after all, to have a significant impact on particular strands of Anglo-American human geography. Thrift himself offers four principal examples, which for him, signify new openings for re-vitalising politics and what counts as political. First, given that our environmental surroundings are now understood to be suffused with 'regimes of feeling', or 'velvet dictatorships' that may come from state or corporate institutions with particular agendas (Thrift, 2004, p 68), it will be necessary to take a more behavioural approach to the education of citizens. As Thrift puts it, to develop '*skilful comportment* which allows us to be open to receiving new affectively charged disclosive spaces' (2004, p 70, original emphasis). This may include body training and Buddhist-informed educational styles 'which use the half-second delay to act into a situation with good judgement' (Thrift, 2004, p 70). It is notable that mindfulness training is already being delivered within schools and workplaces in the UK. The political effects of these forms of comportment have not been investigated, and as Chapters Four and Five explore, there is no reason to suppose that an emphasis on affective behaviours will be as necessarily politically progressive as Thrift suggests. That said, it is also not clear how one might consciously *use* a half-second delay, if, as Thrift asserts, one is *not* in conscious control of one's actions. The second example of a re-vitalised politics is that of psychoanalytical models of affect that take as their method *reparative knowing*, forms of political mobilisation (he identifies post-colonial, sexual and ethnic identity activists) which themselves act on 'perceptual systems by working on associating affective response in both thought and extension' (Thrift, 2004, p 70).

The third direction derived from this politics of affect is split into two parts. First it refers to *tending*, which is described as widening 'the potential number of interactions a living thing can enter into, to widen the margin of "play".' This seems to confer some kind of symbiotic relationship between humans and their ecologies (although these are no longer separate entities), and refers to a 'belonging-together of processually unique and divergent forms of life' (Massumi, cited in Thrift, 2004, p 71). Second, the purpose of this biological-cultural symbiosis is to specifically develop new forms of *neuropolitics*, drawing on William Connolly's book of the same name. It is posited that because contemporary life is suffused with subtle and subliminal 'biological-cum-cultural gymnastics' in the moment between thought and affect (which shapes our decisions and actions), we have become vulnerable to a threatening new fertile 'field of persuasion and manipulation' (Thrift, 2004, p 71). We therefore need to develop a 'microbiopolitics which understands the insufficiency of argument'; in other words, it is no longer enough to engage in apparently rational political argumentation, since our thoughts are always already determined within a pre-rational (or post-rational) domain. Instead, we must pay attention to *how the self is cultivated*, we must be more generous to the world, and we must consider how space and time are re-configured, particularly by new technologies.

The fourth, perhaps most unusual, political direction involves focusing attention on a neo-Darwinian conception of the 'face and faciality' (Thrift, 2004, p 72) as the most important feature associated with being human. In developing this last aspect, Thrift discusses *The Passions*, a piece of video art work by Bill Viola, that depicts faces which at first glance appear to viewers be completely still. Slowly, the faces reveal tiny, slow movements which depict highly staged facial expressions and emotional intensities across an extended field of time. For Thrift, these works also narrate our long-running cultural engagements with the affective face, by containing narrative traces of images of Christ, scientific approaches to physiognomy (reading a person's character from the appearance of their face), and perception, technological advancements in moving image and facial recognition systems and modern press reporting (Thrift, 2004, p 73). What Thrift draws out from this study of Viola's work is the idea that the face is a source of affect which produces meanings redolent of the cultural history of emotions, and reveals how the self has been socialised by minute 'mimetic' practices over centuries. So, too, the tactic of slowing down the videos exposes for Thrift the way in which modern life, and specifically Western culture, is at work in the minute gestures of everyday life, opening up a political space to investigate 'how we are made to be/be connected' (Thrift, 2004, p 74).

As is perhaps by now self-evident, this set of ideas makes some grand and complicated claims to 'know' life, but their political ramifications and specific spatial qualities are often not explicitly set out. However, there is no doubt that its main protagonists are genuinely seeking to get to grips with a vastly complex series of developments in philosophy, social theory and the biological sciences in order to open up new avenues of enquiry into the arguably neglected arena of emotions, affects and practices. They are interested in considering the political and ethical stakes of this approach and its potential to 'think anew'. Given the influential nature of this neural turn on the discipline of human geography, it is important to discern the political significance of non-representational theory in its approach to human life and its potential insights for a geographical understanding of brain culture. Put crudely, the non-representational approach seems to propose that: (1) we have never been entirely 'human' in the sense of being the rational authors of our own destinies in a world in which we are separate from our environments, non-human things and animals; (2) most cultural and social theories are now inadequate for the task of attending to what it means to be human, since they continue wrongly to hold on to a view of subjectivity as socially constructed rather than biologically full of possibilities; (3) social science will only be worthwhile if it forms an allegiance with the performing arts in order to improve its capacities for 'wonder', for 'renewal' and for imagining alternative worlds – this will involve paying closer attention to a wide range of affects and sensations, stepping away from a traditional emphasis on signs and significations, and appreciating that cultural practices shape people rather than vice versa (Thrift, 2007, p 12); and (4) affective life itself is increasingly an object of governance, but that by attending to affect, research can open up new generative political spaces which evade order and control (B. Anderson, 2012, p 29).

Non-representational theory and its related spatial politics of affect have not entered the discipline of human geography without significant criticism. Benedikt Korf (2008) takes particular issue with the neural turn underpinning non-representational theory in geography. For Korf, the non-representational approach is based on a highly deterministic attack on the rational modernist subject, which relies on the naïve naturalism of presuming that physical brain processes *cause* human action. This is relevant here insofar as 'brain culture' and its associated policies and practices rely heavily on the claim that we can no longer be understood as rational actors. At the same time, the atomistic, wholly *rational* modern self has already been problematised from a wide variety of approaches to human geography. As Korf notes, geographers have emphasised the non-rational and emotional aspects of human behaviour, have considered the relational ethics of human and non-human actors in a network of interrelations, and have explored the role of the 'other' in ascribing ethical responsibility (Korf, 2008, p 716). The key difference between these varied takes on the human subject and human agency is that, unlike the non-representational approach, they consider subjectivity to be *contingent but not determined* (Korf, 2008, p 716). This is clearly at odds with the explanatory appeal to Libet's work made by Thrift.

The apparent reliance of non-representational geography on science for its strong political claims has been challenged by Constantina Papoulias and Felicity Callard (2010). In their detailed exploration of the relationship between cultural theory and science, they show how non-representational geography (and the affective turn in cultural theory) is enamoured with a partial (and in their view, mistranslated) vision of neurobiology. These neurobiological claims to know human subjectivity, will and consciousness have important ramifications for understanding political agency and the very possibility of social change. Non-representational geographies are not without merit for Papoulias and Callard, who note how they have set out to address the relative neglect of the body and of the material world in social and cultural theory. They concede, too, that for many, social constructivist approaches to identity have been too pre-occupied with the power effects of discourse and language, and a 'reductive' explanation of how social structures underpin human subjectivity. But the very particular version of biology favoured by affect theorists and non-representational geographers, they argue, is equally partial and reductive. It is an account of the body which privileges the affective experience of the body over and above 'the overwriting of the body through subjectivity and personal history' (Papoulias and Callard, 2010, p 34).

The concluding chapter examines in more detail the emerging (re-) alignment between the biological and social sciences exemplified by non-representational geographies, and the way some old hierarchies between scientific fact and 'mere' cultural opinion and personal biography are reproduced in this way. For the moment, it is sufficient to outline the way in which non-representational theory in particular has been anchored to a specifically imagined biological account of personhood and thinking. Papoulias and Callard mention, for instance, Thrift and others' references to the work of neuroscientist Antonio Damasio, who has developed a particularly philosophical stance to his studies of the role of emotions in thinking. He has been influenced by the writings of 17th-century Dutch philosopher Baruch Spinoza, who questioned Descartes' mind–body dualism. Damasio is invoked by Thrift because of his attention to the bodily, sensorial and emotional *precursors* to cognition (Thrift, 2008, p 59). For Thrift, these precursors are instructive precisely because they open up a temporal gap between embodied action and reflective thought within which is held the possibility of dynamic change, creative force and political invention. On these terms, the body is no longer simply a canvas onto which social norms are inscribed, through which social structures are internalised. Thrift's (2008, p 61) purpose is therefore to highlight 'that blink between action and performance in which the world is pre-set by biological and cultural instincts which bear both extraordinary genealogical freight – and a potential for potentiality.' But as Papoulias and Callard contend (2010, p 41), Damasio's biology is far from 'emergent' and radically open in this manner. Rather, for Damasio, the emotional functions of the brain are set in train along evolutionary timeframes which shape these automatic pre-adapted neural functions in particular kinds of perceptual environments (pre-historic humans running away from their predators). There is a latent determinism in this biological framework which is potentially incompatible with our democratic assumptions of free will.

It would be unfair to suggest that non-representational theory has been wholly driven by its neural turn, or is entirely dependent on the deconstruction of cultural subjectivity which it takes from the insights of specific neuroscientists. However, as the critique offered by Papoulias and Callard (2010) indicates, this turn to affect in cultural geography has been somewhat eager to incorporate neuroscientific knowledge in a decidedly one-way transdisciplinary conversation. It is therefore essential to interrogate Thrift's use of Libet's experiments as a founding principal for his assertion of humanity's 'lateness' for consciousness and the brain's apparent capacity to determine thought, not least because of the ambitious political project that has been built up around these claims. It is Raymond Tallis, who we encountered at length in Chapter One, who again provides us with the most trenchant assessment of Libet's conceptual errors. And it is precisely an inattention to *context* that Tallis identifies as responsible for such errors – making it all the more surprising that geographers might compound them. One of Tallis's main observations on Libet's work is that his denial of free will was made 'on the basis of experiments that are *uprooted from the contexts that make sense of actions*: or, more precisely, reduce actions to movements' (2011, p 247, emphasis added). As is so often the trouble with translating research in the neuroscientific laboratory to implications in the complex social world, Tallis argues that Libet's reduction of 'action' to the request to make a hand movement within the specific context of the laboratory is itself the *source* of Libet's troubling denial of human consciousness and intentionality, rather than the *outcome* of scientific experimentation which can be taken as truth. By focusing the recording equipment (EEG) and experimental design on the moment of hand movement, Libet's research ignored all the action that preceded it, including the research participants' decisions to take part in the experiment, get up in the morning, agree to the instructions of the research team and then to move their hand (Tallis, 2011, p 248). 'It is no surprise' observes Tallis 'that we cannot find free will in this isolated moment in a laboratory, if we treat it as an isolated moment' (2011, p 250). To focus on the decision-making moment related to such a small task as moving the hand, and thereafter to extrapolate a whole political project from the observation of a delay between action and that thought which is discernable by the monitoring equipment of neuroscientists, thus begins to look like an imaginative leap too far.

At this point, we have covered some complex, discipline-specific ground, so it is worth drawing out the significance of non-representational approaches to the explicitly political worlds of citizenship and governance, which are the focus of this book. We can also pause here to consider some of the more explicitly spatial aspects of this work. In sum, the non-representational style of thinking has been a much lauded strain within the discipline of human geography since the late 1990s. It was posited partly as a rejoinder to cultural and social theories which outlined the social construction of human subjectivity and the internalisation of all manner of cultural, ideological, social, political and economic identities, and partly as a corrective to what was seen as an inadequate account of the embodied, material, emotional and affective richness of human being-in-the-world. To this end, non-representational thinkers in geography have drawn on particular insights from the brain sciences, most notably relating the pre-cognitive aspects of thinking and the sensory determinants of action. In so doing, they have made political claims about the possibility of thinking otherwise – in more non-rational ways, beyond and outside of the systems of representation – as a challenge to the broad status quo of the social sciences and cultural theorists. As I have outlined, several commentators have taken the non-representational approach to task for its troubling attitude towards free will and its resultant foreclosure of the ethical human agent, its ambiguous account of the possibilities of subliminal political manipulation through affective architectures, and its dismissive stance towards the constitutive power of social and cultural representations in constituting subjectivity.

In Thrift's earlier elucidations of non-representational theory, the implicit political promise of this style of thinking was not followed by any specific lessons for citizenship and governance. Instead, we are encouraged to pay more attention to 'body practices' (for example, courses in body language, corporate training programmes on awareness and self-presentation), 'mystical practices' (for example, contemplation, prayer, meditation, New Age religions), 'ritual practices' (music, dance, theatre, mime, art, postures, repetitive movements and memorising), and 'body therapy' (approaches to dance, music, massage therapy and 'body-mind centring', which use movement in order to rework emotions) (Thrift, 2008, pp 65-6). The connection between these often non-ordinary practices and everyday relations of power has often been unclear in this work, but Amin and Thrift's most recent articulation of the political ramifications of this approach, in their book, *Arts of the political* (2013), has set out a more specific programme for a non-representational politics. Recognising the current policy enthusiasm for scientific approaches aimed at nudging citizens by intervening in the momentary spaces between action and cognition, Amin and Thrift (2013, p 14) caution that there is something very suspicious about 'a pre-political realm that depends on the black arts of what we might call pre-meditation.' They argue for a new kind of political arts which takes seriously the emotional planes of experience, non-cognitive and automated decisions, non-human agency and aesthetics. They contend that those on the right wing of politics have been historically effective at these political arts while the left need to develop their skills and capacities for 'invention, organization and affect' in order to make new worlds (2013, p 10). *Invention* refers to the ability to bring together

58

new communities of interest (or actors, spaces, political styles, senses and affective fields of action), making them 'see and long for a different future'. *Organisation* describes how progressive political causes require processes and bureaucracies through which to be advanced, carried out and achieved (Amin and Thrift, 2013, p 12), and *affect*, as already noted, describes momentary, impulsive, contagious registers of political decision-making that cannot be understood by an appeal to rational, deliberative thought (Amin and Thrift, 2013, p 14).

The non-representational approach, while focusing on the very possibility of politics, has paid relatively little attention to questions of citizenship formation and governance practices – notwithstanding an enthusiasm for developing an anthropological approach to the study of bureaucratic organisations such as the EU (Amin and Thrift, 2013, p 116), an approach already well-established within the social sciences. It is perhaps no wonder then that the citizen posed here, much like the human subject of brain culture, is diminished to a figure drowned out by its affective capacities and pre-figurative neural drivers, fooled by its own post-hoc rationalisations for action. This neurally inflected citizen, determined by their biology to act in sometimes more but most often less-than-rational ways, is susceptible to affective forms of manipulation. This leads Amin and Thrift to argue that left politics must necessarily similarly adopt 'political psychotechnics' (2013, p 172) if it is to have any chance of political success. The left must draw down databases which track citizen behaviour, relations, affinities, identity and emotional states, and use this knowledge in developing political campaigns based on mood manipulation (Amin and Thrift, 2013, p 173). These techniques, according to Amin and Thrift, have become the mainstay of 'an increasingly insistent media-marketing logic pushed by a motely crew of consultants' (Amin and Thrift, 2013, p 171), and the left is at risk of being left behind. The rise of psychological forms of governance targeted at the 'predictably irrational' citizen certainly brings these concerns to the fore, particularly where a hierarchy of rationality is produced, with certain social groups ascribed more capacity for reflection than others (Jones et al, 2013). And yet non-representational geography does not offer any clear guidance on how we might critically assess this motley crew of mood messengers, marketers (and one might add) affective architects, behaviour change experts, neuroeducators and psychological consultants and trainers.

Critics have suggested that the political implications of the non-representational approach are necessarily self-defeating because of (I crudely simplify here) a persistent tendency to prioritise time over space – a move that may puzzle geographers. Barnett (2013, pp 13-14) identifies in this work two contrasting 'logical geographies of action' – logics shared by the neurobiologically influenced political scientists encountered in Chapter One, who deplored the spatial metaphor of decision-making as dependent on context. The logic given most value within non-representational approaches to geography is famously the temporal or vertical plane – in which action preceded thought and a space opens up for either progressive change or nefarious subliminal manipulation. The second logic, on the horizontal plane, relates to the sense in which the human subject has become relational – that the boundaries of the human skin or skull no longer suffice in making up people and making up action. A whole series of non-human, technological, discursive and material actors is now implicated in what Thrift (2008, p 85) has termed 'a distribution of subjectivity in a population of actors.'

Might we therefore draw some insight from a closer examination of this horizontal plane of human action, promising as it might be for the overall concern in this book to re-situate subjectivity in particular contexts? Thrift's consideration of *where* decisions take place offers some insight into the importance of *locating* subjectivity (2008, p 83). Once we acknowledge that the human subject is no longer as internally coherent as had been supposed prior to the post-structuralist philosophical turn (a turn by no means universally acknowledged), we must necessarily seek new ways of understanding where to find the subject. Thrift explores three potential candidates which offer some useful pointers. First, he considers the *brain*, but determines this to be more of a processer than a source of agency ('as much a transmitter and receiver as a fixed node'; see Thrift, 2008, p 84). This does not seem to be in keeping with his analysis of the political potential of the brain's half-second delay. Second, he highlights the *intersubjective interactions* between people, but finds this too dismissive of non-human agents. Third, he specifies the *unconscious*, but judges this to be an inadequate account of the mutually constitutive experience of both the unconscious and conscious. He therefore settles on a notion of subjectivity informed by the work of the 19th-century French sociologist and psychologist Gabriel Tarde, who focused his attentions on the 'sites at which behaviour was modified, that is with the moment, the location, and the mechanism through which difference or invention was produced' (Thrift, 2008, p 84).

The definition of subjectivity therefore proposed is: 'lines or fields of concernful and affecting interaction taking place in time', and the widespread, individual, recurring (and rather spectacle-like) concern for the effigy of Princess Diana is given as an example as an event which can be understood as both a national zeitgeist and a highly individual felt experience (Thrift, 2008, p 85). In this account, it is still possible to be a person, but the boundaries between people are blurred, and subjectivity now 'flows through' people in ways over which the person has limited control (although modifications are possible). What Thrift seems to be getting at here is similar to the distributed sense of cognition promoted by Tallis, Choudhury and Slaby and the philosophers of mind (Hurley, Noë) encountered in Chapter One. In this sense, the non-representational style of thinking could act as a valuable challenge to the tendency within brain culture to locate human consciousness and, thereby, human subjectivity, narrowly in the brain. If we can rather selectively take the spatial lessons of the non-representational approach rather than its adherence to a rather deterministic set of lessons from neuroscience which privilege the temporal plane of human cognition, then this set of complex ideas may have a lot to offer an analysis of the geographies of brain culture. It can open up our conception of the causes of human action to a new geography of subjectivity in which many actors (including humans, animals and objects, which are said to have their own animacy) are implicated in shaping behaviour. In this account, space is said to have 'its own push', playing an active role in constituting not just human action, but human beingness itself. One doesn't have to accept the idea that things like space, animals and objects have human-like consciousness and intentionality (and that we as humans, by contrast, are not as conscious as we thought we were) in order to appreciate that they make up the contexts in which possible human actions are shaped. But if we are to draw out any substantive lessons from this geographical approach, then we must re-focus our analysis on the actual everyday contexts in which subjectivity might flow through people in specific ways. We must not dismiss intersubjective social relations, nor downplay the historical and geographical specificity of cultural productions of personhood. Furthermore, we must turn our attentions to the often mundane practices of citizen formation, public policy and governance techniques rather than limiting analysis to often abstract and highly staged, artistic, spectacular and theatrical practice – as is sometimes the case for non-representational geography.

Contextual rationalities

A third set of ideas within human geography goes a long way to addressing some of the traditional spatial naïveté of behavioural geography and environmental psychology, and tackles the relationship between space and thinking in a more everyday and contextually specific manner than is proposed by a non-representational approach. Foucault's work on governmentality has become particularly prominent in geographers' accounts of the arts of governing the self, the citizen and populations, and thus provides a useful complement to critical approaches to the history and practice of neuroscience outlined in Chapter One. Since at least the 1990s, geographers have drawn inspiration from Foucault in investigating the intersections of knowledge, space and power (Philo, 1992; Hannah, 2000; Elden, 2001), the governing of 'life itself' (B. Anderson, 2012; Kraftl, 2013), rationalities of governance (Larner, 2000; Barnett, 2001; Huxley, 2006; Cadman, 2010; Jones et al, 2011b), and the technologies of the self (Barnett et al, 2005). In this section, I examine just one aspect of this work which explores understandings of the constitutive role of *rationalities* to an understanding of the political significance of brain culture *in context*. I use the term 'context' to denote an emphasis not just on space, but on specific spaces, and not just on material physical phenomena, but on discursive terrains of action. Rationalities, in turn, refer to the truths, thoughts or knowledges that denote the aims of government (Huxley, 2006, p 772). Huxley's work, among others, acknowledges the limits of fetishising 'practices' as if they take place in a social, political or spatial vacuum, and instead turns our attention to a 'regime of rationality' which can be said to 'found, justify and provide reasons and principles for these ways of doing things' (Foucault, 1991 cited in Huxley, 2006, p 771).

An enlightening review of Foucauldian approaches to geography offered by Chris Philo is helpful here. He advises that Foucauldian approaches are not as distinct from those of the non-representational theorists who have been quite so dissatisfied with the main tenets of Foucault's work – because of this apparent prioritisation of discourse (and social construction) over embodied practice (Philo, 2012, p 498). Philo clearly demonstrates how Foucault's writings have an ongoing contribution for geographical understandings of vital bodies (including, one might add, neurobiological accounts of brains) and populations. For Philo, Foucauldian approaches to geography are not as ignorant of fleshy materialities, vital bodies, non-textual and resistive practices as their accusers would have it (Philo, 2012, p 499). Rather, his original work is full of accounts of *both* the pre-discursive body *and* power struggles over its cultural inscriptions.

To summarise, an understanding of geography informed by Foucault can shed light on human subjectivity at two scales of explanation. It illuminates the always incomplete and sometimes unsuccessful orchestration of spatial arrangements in the securing of power over bodies (*anatomo-politics*). It also brings to the fore those governing techniques aimed at managing whole populations through a biological mode of address targeted at societies as species, through statistical adherence to social norms and through the elimination of deviant bodies (*biopolitics*) (Philo, 2012, p 505). The point is not that the historical and contextually specific forms of power outlined by Foucault render the material body powerless to the discursive tropes which overlay it, but conversely, that the body marks an excessive and ultimately untameable lifeforce which carries with it the potential to 'spill out' (Philo, 2012, p 506) – and thus resist political domination. In both the non-representational and Foucauldian approaches to geography, therefore, the biophysical body is both the site of political manipulation and the source of political agency. In the case of the former, this site is the 'half-second delay' between action and thought, and for the latter, it is the 'lively bodies and populations over which (dominating) power is to be exerted, and whose pre-discursive liveliness is often too much for the powers-that-be' (Philo, 2012, p 510).

Hence, when we look at the proliferation of brain culture, as a discourse by which we come to know ourselves through the rhetorical lens of the neurosciences in such spheres as architecture, education or workplace training, we need to look at both discourse and embodied practice. Discursive forms of knowledge, institutions and texts give meaning to the human brain *in situ*, while embodied practices give us insights into everyday experience. Together these make up the social, economic, cultural and political contexts which render certain actions possible. Understanding various (contingent and often incomplete) attempts to shape spaces, or to 'create milieux that would induce and maintain specific and differentiated forms of conduct', as Huxley echoes Foucault in explaining (2006, p 773), is therefore central to our concern here to outline the citizenship and governance implications of brain culture. This is not to say that bodies and populations cannot and do not resist attempts to programmatically govern them in this way, as the preceding discussion of Foucauldian geography insists. And yet it would be naïve to suggest that governments do not or should not attempt to arrange spaces, orchestrate interactions, manage populations and shape the very biological *and* political realities of citizens (their health, wealth and happiness). Huxley's own interest is in the history of urban design and various spatial arrangements which attempt to shape behaviour through the adoption of all manner of statistical, medical, architectural and design knowledges, as well as philosophical and spiritual ideas – in the service of making virtuous city spaces in which order, health, wealth, happiness and the creative evolution of humanity may flourish. She terms these 'vitalist spatial rationalities' (Huxley, 2006, p 780), again indicating the embodied and yet also discursive forms of governmental reason deployed here. Her work relates to city planning and therefore a quite literal understanding of spatial context particularly relevant to the analysis of urban design and neuroarchitecture developed in Chapter Three.

But here I want to outline how a wider appreciation and rather more scaled-up rendition of the 'geo-historical' contexts of human activity might shed new light on both the broader political rationalities behind brain culture and its unintended consequences.

Attending to the 'geo-historical' contexts of human activity means conceiving of the human as simultaneously a discursive subject and a natural, vital body. It is a social construction subject to relations of power and a *real thing* subject to bodily limitations and potentials. Yet the circularity effect of brain culture – the way in which knowledges about the brain become the basis of social intervention, public policy, popular culture and everyday practice, and thus feed back into our understandings of the 'self' which is the object of so much neuroscientific endeavour – means that the separation of the discursive 'me' and the embodied 'I' is indeed not possible. That is to say that the biologically rendered, pre-cognitive human subject of neuroscientific experimentation, behavioural geography, environmental psychology and of non-representational theory (to identify one continuity between these distinct schools of thought) is inaccessible to human thought, and should therefore be considered a *real fiction*. However, the subject is not, from this perspective, 'dead', but re-animated and re-situated in their specific time and space (more like the account of distributed cognition mentioned earlier). It is made, but not 'made up'. This much more tentatively socially constructionist view does not necessarily confer an ignorance of the body, its materialities and its bio-physical realities. Instead, it is context-sensitive in its attempt to relocate these bodies and the production of biological narratives about them within specific regimes of social and political rationality. In so doing, we can further explore the spaces and practices through which these subjects are made governable, and can make more visible the specific work that goes into telling specific stories about being human (bodies) in several different spheres.

It is useful at this juncture to elaborate on a critical distinction between subjectivity and human action which have been somewhat blended together both in the preceding discussion and in the geographical literature outlined above. As Häkli and Kallio (2013) have pointed out, in the rush to declare the death of the modern subject (whether by poststructuralist philosophers, in their argument, or by neuroscientists' endeavours to locate the neurophysical source of consciousness), questions of political agency and the everyday enactment of politics have been marginalised. Human geography thus appears divided between those who pay attention to the immanent, creative and open-ended performances of subjectivity (such as is the case for non-representational approaches) and those whose concern is to understand the social and political struggle over identity politics, subjectification and recognition (by implication, social constructionists and discourse theorists such as Foucault). In the analysis offered by Häkli and Kallio (2013, p 185), neither of these positions can properly account for subjectivity and political agency. Such false distinctions, they assert, diminish geography's potential for re-contextualising that political agency in specific times and spaces – or in the terms developed in this book, in geo-historical epochs. In offering a more pragmatist account of political agency, Häkli and Kallio therefore propose that we must pay closer attention to the dynamic interaction between the 'I' (self) and 'me' (identity) which is unique to the human subject. The 'I' refers to the immanent experiences of the reflecting subject, whose object of reflection is the socially constituted and contextually specific 'me' – an identity that emerges in response to discursive social categories, moral norms, embodied experiences and the values associated with attachment to or membership of particular communities. It is therefore imperative to take account of the specific context in which the 'me' is socially constituted and constrained, while recognising the agency of the 'I' in ensuring that action can unfold in creative, unpredictable, contradictory and resistant ways. Their approach is an original attempt to outline a specific political geography of the discursive *and* embodied citizen, an analysis in which context takes centre stage. It is therefore in the context of the 'polis' as a space produced by the intractable conflict engendered by living together socially with our competing desires, interests and goals that this dynamic intra- and inter-personal struggle takes on significance:

> Our political agency, then, begins when we enter into social relations that animate the dialogue between "I" and "me" and our (significant) others. Consequently, our agency in the polis is marked less by the battle between some authentic inner self and demands coming from the society than by *how we relate subjectively to situations*, events and positions offered to us in the course of our lives. This seemingly

subtle move is important because it shifts the relationality of "the political" from within the individual into the social world that the embodied individual encounters in multiple different subject positions, averting, accepting or altering them through individual or concerted action. (Häkli and Kallio, 2013, p 11, emphasis added)

What Häkli and Kallio's emphasis on context does for us is describe the centrality of geography to any realistic account of political action. While the behavioural geographers, environmental psychologists and non-representational geographers discussed here have made allusions to context in the various ways noted above, they all tend to prioritise the political agent as existing 'inside' the person (the individual mind, psychological traits or neurobiological drivers of action). As such, as claims to know the world and as forms of analysis, they are all, to a certain extent, co-implicated in the production of brain culture, an account of the world that has the brain as the source of that world. Most curiously perhaps, in their shared neglect of or resistance to social constructionism and in their prioritisation of the vital, physical and material properties of the human brain and body, what these approaches have done is downplay that everyday, worldly context in which political and social subjectivities are formed and re-formed in simultaneously dynamic, contingent, discursive, embodied and situated ways. These formations are subject to political contestation, deliberative argument and democratic process in a manner not shared by a neurobiological basis for human action.

Replacing neurogeography with a geography of brain culture

When neuroscientists claim to know something of human experience, there is often little recognition of the importance of context in shaping that experience, and certainly no shared conceptualisation of context itself. Moreover, where brain culture has influenced policy and practice, it is more often than not through an insistence on the intra-personal and intra-psychic conflict between the emotional/rational brain, the automatic/reflexive self and the systematic neural errors that deceive us. This justifies all manner of practical and policy interventions. These interventions undermine the significance of context – understood in social terms of a space produced through *inter*-subjective and political struggle – in shaping human subjectivity. The attendant production of putatively universal 'neurocitzens' to be governed through brain culture bypasses the very same political struggle which is temporarily instantiated in the figure of the citizen-subject which is always in flux (Isin, 2009). By paying attention to contextual rationalities, in contrast, we can describe the way in which brain culture itself constitutes people and re-invents citizens through providing the rhetorical impetus to govern the self, govern through the brain, and through the practical and discursive processes of normalisation and subjectification outlined in Chapter One. A concern for geo-historical contexts also goes some way to addressing the neurosciences' aforementioned problem with scale, and its narrow focus on the immediate ecology of situations.

While human geography might seem a reasonable starting point for theorising context and for understanding the brain *in situ*, the discipline itself has been influenced by brain culture in many guises, from its appeal to psychological accounts of human–environment interaction to the neural turn and 'affective architectures' of non-representational theory. This focus on the neuromolecular, mental process, the brain, the mind or behaviour necessarily provides a blinkered view of the determinants of human action. The decision-making sciences may well be attentive to the genetic determinants of conscious thought, the responses of the brain to environmental stimuli, conditioning, even the influence of particular 'choice architectures' which guide our decisions, or the affective architectures that foreclose rational action. But they do not have the conceptual apparatus or the methodological means required to understand the person as much more than a misjudgement-prone, automated and yet (in a limited fashion) 'choosing' subject. The approach taken in this book is therefore to foreground the contextual rationalities within which the neurocitizen is governed within actually existing brain culture. In so doing it offers a critical reading of how the person as both a political subject and agent is situated in history and in geography. It provides a modest geographical account which – although always partial – challenges the rhetorical certainty by which the behavioural, psychological and neurosciences can claim to know and seek to shape human action.

An example of a context-sensitive geographical form of explanation may be helpful here. The contemporary problem of obesity, for instance, has often been described in psychological terms (Cooper and Fairburn, 2001). People partake in habitual behaviours of over-eating, or eating high-calorie foods, and as such they are said to need cognitive behavioural therapies, community weight-loss groups, or drugs and surgery in order to control their eating behaviours or bypass their appetites altogether. Or it is understood in neurobiological terms. Obesity is like drug addiction: food activates the dopamine reward system in the brain, and where there are deficits in this system, animals may over-eat in order to compensate (Geiger et al, 2009). Indeed it may also be understood in evolutionary terms – our modern-day (Western) lifestyle is simply not in keeping with our biological make-up, adapted as it is for food scarcity rather than abundance. Such accounts of obesity are not ignorant of context – the social context of both eating and dieting is often well understood in terms of either 'obeseogenic' environments, or in relation to the behaviour-shaping influence of social norms (Robinson and Higgs, 2013). However, by contrast, relatively little attention is paid in these accounts to (1) the wider culturally specific context in which being fat has become an undesirable subject-position (see Colls and Evans, 2009); (2) the social context in which food has been reworked as a fuel for maintaining the working body; and (3) the political and economic context that has shaped not only the availability of (un)healthy foods (the obesogenic environments), but also the ingredients that make up processed foods, as well as the advertisements that popularise high-calorie, high-fat and high-sugar eating, and the regulatory contexts that have failed to address any of these issues in the face of overt lobbying by the powerful multinational food and drinks industries (Moodie et al, 2013). We can therefore begin to appreciate how an expansive notion of the geo-historical contexts of human activity proposes some quite different rationales for behaviour beyond biological, psychological or neurological explanation. Such rationales suggest substantially different programmes for governmental intervention aimed at social, cultural and environmental change rather than promoting incremental modifications of behaviour. Further examples of the potential limitations of 'neurogeographical' explanation could be taken from some accounts of human-induced climate change offered by environmental psychologists (those which posit individual behaviour as the source of and solution to climate change), or from neuroeconomic explanations of compulsive spending, financial over-reach and myopia in the global financial crisis (as explained in Clark, 2011).

Critical and contextual accounts are of course also available from outside geography. Many of those scholars introduced in Chapter One have offered understandings of a spatially distributed rationality of brain culture. They have started to develop accounts of human activity which are more sensitive to the unequal distribution of power within specific contexts. This can potentially be understood in at least two ways: first, that the power of individuals to change, for example, their environmental, financial or eating behaviours, is clearly much less than the culture-shaping power of multinational corporations or the regulatory power of governments. Even if we are to think of cognition and subjectivity themselves to be distributed, this does not mean we can do away with inequalities of power. In a second sense, we must understand brain culture itself to be a geo-historical rationality reliant on, as Abi-Rached and Rose (2013, p 20) contend, 'the intense capitalization of scientific knowledge' in the contemporary 'neuroeconomy'. Their critical account of the institutional dynamics, funding and commercial–state partnerships for neuroscientific research helps us to situate this culture. This context-sensitivity is useful in better appreciating the way in which brain culture shapes the frames of reference in which subjects see themselves, and goes some way to explaining how neuroscience has gained sufficient rhetorical status to set agendas for policy in practice in a range of real contexts. Yet we must come back again to one of the main blind-spots of neuroscience and unintended consequences of brain culture. This is that the decision-making sciences and neurosciences cannot adequately account for their own role in the shaping of the people whom they seek to understand, for there is nothing natural about their (collectively produced, well-funded, institutionalised, technologically enabled) knowledge. As Choudhury and Slaby (2012, p 3) have pointed out, the neurosciences have no means to comprehend their own role in shaping the very *subjectivity* which they purport to *objectively* and scientifically understand. Such is the inevitable and problematic circularity of the brain world.

The notion of a 'critical neuroscience' has been forwarded as a means to question biological claims to know human nature or subjectivity (Slaby, 2010). To this end, since 2009, a group of scholars originally centred in Berlin have begun to re-contextualise neouroscientific knowledge in ways that can greatly strengthen a geographical analysis of brain culture, policy and practice. Critical neuroscience develops the insights of science and technology studies in order to examine the production of scientific facts enabled through an attention to the social and political make-up of scientific culture. This emerging research programme has started to trace the journeys of 'brain facts' from laboratory to application, at the same time as attempting to introduce sociological, philosophical and historical perspectives into neuroscientific research programmes. These scholars are forthright in rejecting any separation between perception and action, and instead focus on 'enaction', building on those embodied philosophies of mind encountered in Chapter One. Their approach is highly context-sensitive in the way in which it aims to foreground life experiences and social structures, and re-embed the brain in both its bodily dimensions and the physical and social environment, understood as the person in the ecological system, institutional environments and even cultural triggers of neural processes (Choudhury and Slaby, 2012, p 11). But in distinction to social and cultural neuroscientists, *critical* neuroscientists consider the social as irreducible to the brain or to neuroscientific categorisation. Their approach points to a new way of doing neuroscience and social science, proposals that form the backdrop of this book.

In forwarding a geographical analysis of brain culture in context, this chapter has sought to establish a framework that is sufficient to deal with the vital materialities of the brain and body as well as the discursive formations through which people are constituted as neural subjects. It has examined several geographical perspectives on the question of what the 'environment' of brain–environment interactions, the affective 'architectures' of human behaviour and the 'contextual rationalities' of human subjectivity and political agency might look like. It has also explored some of the limitations of existing geographical approaches and their tendency to accede, and in some cases defer to, explanations of human behaviour and personhood offered by the behavioural, psychological and neurobiological sciences. It draws on Foucauldian perspectives on geography in order to argue for a focus on the contextual rationalities of brain culture. Despite the apparent falling out of fashion of Foucault within human geography, and notwithstanding some influential critiques of his work, the value of his analysis for explaining the powerful effects of discursive formations in shaping the conduct of vital human beings in practical terms 'in the *milieu* in which they live' (Foucault, 2003, p 244, cited in Philo, 2012, p 507) should not be underestimated. One does not have to be an expert Foucault scholar to appreciate the critical purchase of this work in understanding the constitutive role of the neurosciences in shaping citizens to be governed within the brain world. By charting the significance of geo-historical context in shaping contemporary brain culture within the UK, we can open up investigation into the spatial politics of governing citizens in the specific fields of policy and practice explored throughout the remaining chapters of the book. To do so provides a cautionary note to modes of behavioural, psychological and neuroscientific governance which increasingly appear above and beyond the realm of everyday politics and ordinary reason.

THREE

Designing cerebral cities

> I contend that architectural design can change our brains
> and behaviour. (Fred Gage, 2009, p xiv)

Introduction

Since the late 1990s, neuroarchitecture has emerged as a paradigm
for integrating new neuroscientific knowledge with architectural
education and practice. Neuroarchitecture seeks to establish applied
approaches to planning, design and engineering which take into
account both the effects of the built environment on brain activity, and
the neurological correlates of space perception – the influence of the
world around us on our embodied cognition and our behaviour. This
chapter introduces some of the main ideas behind neuroarchitecture,
and places it in the broader context of environmental psychology and
urban design disciplines – tracing common paradigms and identifying
key methodological distinctions where necessary. Attempts to design
particular behaviours into spatial environments or buildings, and design
practices that are explicitly aimed at shaping healthier places or better
learning environments, are highlighted.

   The chapter explores some of the practical and conceptual difficulties
of pursuing interdisciplinary investigations into neuroscience and
architecture, and considers the problems associated with translating
neuroscientific research evidence into applications in real places.
The notion of 'spatial behaviour' is investigated, and the way in
which neuroscientists, architects and urban designers conceive of the
'environment' said to shape human behaviour is further examined, in
light of the geographical analysis of contextual rationalities outlined in
the previous chapter. This discussion raises a set of overtly democratic
questions. First, concerning the extent to which neuroarchitecture can
be considered scalable, how valid is the *neuromolecular gaze* in designing
and governing spatial environments for individuals, specific social
groups and whole populations at these different scales? Second, what
is the appropriate role of 'spatial experts' in designing behaviourally
informed built environments through which people are incited to
govern themselves and also be *governed through their brains*? And finally,

to what extent might the new techniques of neuroarchitecture carry with them an intensification of the processes by which to *subjectivise* citizens by enabling the governance of our attention? Put another way, how does the phenomena of neuroarchitecture – manifest in new interdisciplinary research practices and urban design policies – have the potential to re-shape our relationships to ourselves and to our brains in particular spaces?

Brains and buildings

Neuroarchitecture refers to the interdisciplinary endeavours of neuroscientists and architects currently working together in order to discover how our brains respond to architectural experiences. This approach, still in its infancy, has been developed largely through the activities of ANFA, established by the American Institute of Architects in San Diego, California in 2003 (Eberhard, 2009, p vii). John Paul Eberhard, author of *Brain landscape* (2009), has been a leading figure in this venture. During the 1990s he was invited to undertake scientific research on the human experiences of architecture by medical researcher, Jonas Salk, and was to become the founding president of ANFA. The Salk Institute for Biological Studies is an internationally renowned biological research institution, equally well known for its distinctive modernist architecture designed specifically for the practical needs of laboratory researchers in the 1960s by Louis Kahn. Eberhard is an architect with a long and varied career spanning five decades, and in 2005 he became the first architecturally trained member of the Society for Neuroscience, a 42,000-member strong international organisation of scientists and physicians committed to understanding the brain and nervous system (Eberhard, 2009, p 24).[1] Fred Gage (cited at the start of the chapter), past president of the Society for Neuroscience and Salk Institute neuroscientist, gave the keynote address at which ANFA was formed.

By bringing neuroscience to architecture, it was hoped that the scientific shortcomings of architecture could be addressed in order to understand the 'true' impact of the environment on our brains and behaviour. The project of neuroarchitecture sets out to address perceived deficiencies in the evidence offered by architects, who are sometimes said to be too reliant on intuitive conjectures about how buildings affect occupants. While its formalisation through ANFA

---

[1] www.sfn.org/about/mission-and-strategic-plan

may be relatively recent, neuroarchitecture builds on previous research on environmental behaviour from the 1980s, which developed many of the theoretical and methodological precepts used in researching behaviourally inflected spatial and architectural design today. It is also arguably rooted firmly within the history of ecological psychology and the pioneering work of Kurt Lewin and Roger Barker in the 1950s, as outlined in Chapter Two (Zeisel, 2006 [1981], p 80). Neuroarchitecture is concerned with our perceptions of and orientations in space (Eberhard, 2006a, p 11). The contention here is that the brain drives behaviour, and yet there is capacity for the environment to change both the brain and thus our behaviour, through the process of neurogenesis (the growth of brain cells) discovered in the 1990s. It is this behavioural impact of architecture and spatial design that the project of neuroarchitecture seeks to measure and validate. Specifically, neuroarchitecture promises to fulfil the quest for *explanation*, for while it is argued that architects have relied on their intuitions and social scientists on their observations and understandings, neuroscience can give the definitive account of the precise impact of spatial formations on behaviour (Eberhard, 2006a, p 12).

The specific kinds of spaces which are explored within the emergent field of neuroarchitecture largely follow on from research on environment–behaviour interactions, which has been traditionally concerned with therapeutic/assisted living spaces and medical buildings, school, learning and play environments, and workplaces. As a well-known advocate of environmental design and the 'environment–behaviour' (E–B) field, John Zeisel notes that E–B is particularly suited to providing an evidence-based approach to 'constructing environments to improve productivity, quality of life, learning, teamwork, and memory' (Zeisel, 2006, p 16). But now, neuroscientific techniques can be added to the traditional psychological methodological toolkit consisting of observation, surveys, interviewing and investigations of archival drawings and plans. It is now possible to record the impact of spaces on brain physiology, through measures such as 'heart rate, perspiration, and levels of certain hormones and chemicals' which mark emotional responses, as well as fMRI and positron emission tomography (PET) which can 'zero in on the brain regions that could be involved in given responses', while eye-tracking devices can record attention (Eberhard, 2006a, p 12). Zeisel thus recommends the addition of neuroscience to the E–B paradigm, promoting an E/B/N (neuroscience) approach. This focuses on the brain's 'environment system' that plays a crucial role in brain activities relating to the three principal processes of 'memory, orientation, and learning' (2006, p 141). This environmental system is central to cognition in three principal ways.

First, as Zeisel (2006, p 147) notes: '[a]wareness of "place" is critical to the definition of a memory. Physical environment is therefore essential to memory reconstitution.' He explains that our memories are shaped by the place in which we experience something, the place where we are trying to remember that thing, as well as the reason for remembering and the time lag between experience and recall. Place is essential to the accurate working of memory in the brain because we contextualise those memories in reference to a specific time and place. As such, the brain only works within this environment system. There are parallels here with the notion of distributed cognition outlined in the previous chapter, with place and context proving integral to brain functioning.

Second, we orient ourselves in space according to our memories of spatial relationships, wayfinding cues and cognitive maps (Lynch, 1960), and this is essential for finding our way around, evading danger, seeking refuge, and movement more generally. A special place is reserved for the hippocampus as the 'map room' of the brain (Hobson, 1994 cited in Zeisel, 2006, p 148). Landmarks and the relations between them in the physical environment are thus a crucial aspect of brain function.

Finally, the physical properties of the environment are said to be fundamental to learning – the ability to recall what we have learned is dependent on contextual factors such as how we feel, our state of mind and where we learn. It is thought that we ascribe meaning to learning environments that facilitate recall. In this sense, neuroarchitecture has the potential to address some of the perceived shortcomings of the brain sciences in terms of their lack of engagement with brains 'in situ', since the environmental system is understood as a requisite component of neural functioning. However, the environmental system here is confined largely to the micro scale of buildings and room design. Hence, there is a narrowing of the meaning of 'situation' or context that has fundamental consequences for the kinds of behavioural explanations that neuroarchitecture can offer.

Nevertheless, in order to create better environments which 'support our brains' within this environment system, Zeisel (2006, p 143) has argued that understandings of neuroscience are as important as understanding the needs, behaviours, attitudes and opinions of users of space. Neuroarchitecture thus ventures that by understanding not merely *how* but *why* we have particular responses to our environments, we can go beyond meeting users' basic needs (as interpreted by those users), and reassess their needs in light of neuroscientific evidence.

In essence, neuroarchitecture is an invitation to neuroscientists to undertake investigative research into how our brains interact with buildings and urban landscapes (Eberhard, 2009, p 2). To this end, during his tenure as president of ANFA, Eberhard developed a set of hypotheses intended for the research work of postgraduate students and postdoctoral researchers sponsored by ANFA's grants programme.[2] These hypotheses relate to the principal areas of sensation and perception, learning and memory, decision-making, emotion and affect, and movement (Eberhard, 2006b). The settings to which the hypothesis relates include healthcare facilities, therapeutic environments for Alzheimer's patients, classrooms for 5- to 12-year-olds, and sacred and spiritual spaces. Topics to be investigated include: the influence of windows on the healing process; the importance of landmarks in wayfinding inside hospitals; the impact of lighting on the circardian rhythms and sleep patterns of the body in health or education settings; the influence of background noise on cognitive activities in classrooms; the impact of the physical space (availability of private spaces, 'child-sized' spaces, spaces shaped by children themselves, visually contrasting spaces of specific colours) on children's attention; and the relationship between moving through a sacred space and feelings of transcendental religious experience.

In the following two sections, I give two brief examples of how the nascent investigations of neuroarchitecture propose to develop guidelines for designing better learning spaces and spaces for the visually impaired. These examples draw on qualitative interviews with researchers who have been involved with ANFA over the past 10 years: Professor Tom Albright, current president of ANFA and director of the Vision Center Laboratory at the Salk Institute for Biological Studies, and Dr Margaret Tarampi, a research associate of ANFA who is a trained architect, holds a PhD in psychology and has worked alongside neuroscientists in various labs at the Salk Institute while working for ANFA between 2003–07. These two figures have been playing a role in shaping future research agendas in neuroarchitecture, despite having reservations about the term itself. I consider how these research agendas raise questions regarding: (1) the nature of spatial behaviour and competing conceptualisations of the 'environment' which shapes human activity; (2) the problem of translating neuroscientific research evidence into real-world applications; and (3) difficulties in conducting genuinely two-way dialogue between neuroscience and

---

[2]   www.anfarch.org/hay-research-grant-program/

architecture. The final part of the chapter then explores the broader contextual rationalities of emerging insights into neuroarchitecture and environmental psychology as they may be deployed in urban design and governance.

Designing learning spaces

Educational settings are seen as crucial spaces for the application of neuroarchitectual insights, particularly settings that cater for children between the ages of 5-12, because this is regarded as a pivotal period in the development of their brains. Understanding the sensory perceptions, learning patterns and brain development of children could lead to better classroom and school design, as Tom Albright explains:

> 'If you go in a typical classroom – elementary school classroom – on the walls of the classroom – they're sort of chock-a-block with the works of the children. They're cluttered, in other words. So they're statistically complicated and that's an environment that in one context is very reinforcing for the children; that is to say, they have their work displayed on the walls. [But] in a different context in which you're trying to grasp some piece of information, all that stuff on the walls is noise and it may visually interfere with your ability to extract the information that you're being taught; that is, the equations on the chalk board, for example. So the ideal environment would be one in which you can change the statistics in a way that matches the content of the material that's being taught.' (interview, June 2013)

Tom Albright sits on the advisory board of a school design project led by Peter Barrett, Professor of Management in Property and Construction and Fellow of the Institution for Chartered Surveyors, at the University of Salford, Manchester, UK. The HEAD (Holistic Evidence and Design) project has deployed an interdisciplinary approach to investigate the impact of school building design on the learning rates of pupils. The project, funded by the UK's Engineering and Physical Sciences Research Council, has to date used neuroscience and a multi-level modelling framework to understand the impact of six parameters on learning improvement in seven different UK schools: colour, choice, connection, complexity, flexibility and light (Barrett et al, 2013, p 678). The researchers identified the parameters by recording the layout of classrooms including phenomena such as lighting, floor covering, colour and windows. Basic measures of light, noise and CO_2, room height and furniture size were taken, and indicators of sensory comfort (for example, temperature, smell, glare and noise) were established from interviews with classroom teachers. They found that these parameters contributed 25 per cent to the learning rates of pupils, as measured in the National Curriculum levels of attainment achieved in reading, writing and mathematics.

This is by no means the first study to address the question of the impact of building design on learning outcomes, yet it marks a new path in the precise scientific measurement of such impacts, and the development of policy recommendations for school design. The research specifically highlights the need for classrooms that have adequate natural light, good quality artificial lighting, purpose-designed furniture, flexible learning zones that can be adapted for different activities, an appropriate level of visual displays on walls (not too much or too little visual stimulation), and age-appropriate colour schemes (warm for seniors and cool for juniors) (Barrett et al, 2013, p 688).

The architectural design of learning spaces has become an important focus for UK educational policy-makers since the early 2000s, when the New Labour government announced an ambitious new school building programme which would take innovations in the design disciplines to new levels of sophistication and scope (DfES, 2003). The Building Schools for the Future (BSF) programme drew continuously on the perceived need to re-think school design as fit for the 21st century, literally building a transformational vision for the redevelopment, investment and radical refurbishment of the UK's schools infrastructure. This architectural vision has since been extended to UK universities, informed that 21st-century learning requires that 'the design of our learning spaces should become a physical representation of the institution's vision and strategy for learning – responsive, inclusive, and supportive of attainment by all' (JISC, 2006, p 2). The recommendations for both schools and universities are related to the need for flexible learning spaces that could easily be adapted to different activities and future needs, personalised spaces tailored to the needs for individualised learners, game-changing (in terms of challenging existing pedagogies), and technologically sensitive spaces designed with new learning technologies in mind.

In a geographical analysis of the modernist utopian visioning implicit in New Labour's educational reforms and their manifestation in school architectures, Peter Kraftl outlines how a focus on the technical detail of school design, building rules and architectural parameters coincided with a rhetorical framing of BSF as a nation-wide project for ensuring both social inclusion and global educational competitiveness (Kraftl, 2012, p 854). In this way, he identifies a crucial spatial discourse at operation within school building policy and practice, a discourse that posits the local scale of a school building as the solution to a complex myriad of social problems usually condemned as structural, intractable and politically contested (Kraftl, 2012, p 857) – although the subsequent axing of BSF by the coalition government in 2010 strongly hints at the reverse on this latter point. By traversing scales from the micro-parameters of design intervention to the nationally and globally rendered educational and socioeconomic vision, the material, technical and imaginary are brought together in a way that is both intended to shape the learning experiences of the children within the schools, and also to the communities in which those flagship schools are situated and to whom the building will be a highly visible exemplar of New Labour's ideology of social inclusion. Kraftl (2012, pp 862-3) thus notes how minute details 'from procurement processes to financing, from health and safety standards to building regulations that govern the minutiae of windows, insulation, wiring and desk height' sit alongside 'flagship spaces' (atria, halls) aimed to inspire, regenerate and change the behaviour of the community beyond.

For Kraftl, the apparent political neutrality of physical engineering and material architecture is rendered problematic by the way in which buildings are implicated in New Labour's re-visioning of educational futures. Rather than creating neutral places, Kraftl (2012) shows how these new built forms configure new spatial rationalities which are at once material and discursive, embedded within a specific political-economic context which acts on the conduct and perceptions of children, teachers and communities. Such architectural forms, he argues, work to designate the 'child' as a future worker at once responsible for economic stability and competitiveness. They also act to re-frame 'community' as the source of and solution to social deprivation, positing 'lofty' aims that communities can but fail to achieve.

As this discussion of Kraftl's work demonstrates, architectural design understood as a cultural and social practice reveals as much about the political imaginary as it does the technical capacities of buildings. Kraftl's theoretical approach exemplifies the value of a geographical analysis that attends to both the material details of embodied interactions in space and the discursive constructs that are mobilised through such spaces. In so doing, his approach is as sensitive to the biophysical as it is to the relations of power implicated in shaping spaces. Similarly, neuroarchitecture can be understood as a partial political project rather than as a set of novel scientific methods for ascertaining, measuring and ensuring the 'good' design of buildings and spaces. As Chapter Two explored, the scale of explanation offered by neuroscientific insights into architecture has important ramifications in terms of understanding and changing behaviours.

Neuroarchitecture might endeavour to conduct strongly objective science, driven by laboratory practice on perceptual states or animal behaviours. However, when we investigate the spatial and scalar discourses of neuroarchitecture, we can appreciate how the model of the brain in its environmental system eviscerates the meaning of that contextual environment in some unhelpful ways. By focusing on the micro-scale perceptual and sensory environment specifically within the viewpoint of individuals within rooms or buildings, the cultural, social, historical and political parameters of that context are excluded from sight. In a sense, this is to be expected – neuroarchitecture simply asks different questions about learning spaces than social scientists. But there is a danger of developing functionalist accounts of learning that respond to a narrow set of questions around how to use buildings to increase measurable 'learning rates'. Such an approach has two main oversights. First, little attention is paid to what education means, whether it can be predicted and known in this way, and whether a measurement of learning rates adequately captures that which we understand education to be for. And second, learning itself is of course known to be highly context-dependent. The socioeconomic environment and the backgrounds of learners, teachers' performance, the terms of conditions of their employment, and perpetual changes in education policy, are among some of the factors that determine educational outcomes. To attempt to 'control for' these factors in any model of educational attainment appears to be a major omission in investigating how to improve learning. In so doing, learning appears to be reduced to those facets of academic performance that can be accelerated by the right kind of perceptual and visual environment.

It is clear that proponents of neuroarchitecture are well aware of some of these concerns, although they have not necessarily discussed these issues in any great detail. As Tom Albright notes:

> 'The idea is that if we know more about the brain, or as we know more about the brain, we ought to be able to use that information for lots of purposes and the buildings that we inhabit are fundamental to who we are – they make us who we are in many ways and we ought to be able to

modify that in ways that are positive.' *Jessica:* 'But do you have lengthy debates about what positive means in that context? What is a positive change? What are the values that you want to instil in these buildings, for instance?' *Tom:* 'That's a good question and we haven't had a lot of those kinds of discussions. It has come up in the context of school design because there a positive outcome is sort of obvious. We want to be able to improve academic performance.' (interview, June 2013)

Tom Albright is open to such discussions, but is not necessarily worried about how neuroarchitecture might be used in politically partial forms of behavioural change, including those that might not be so straightforwardly positive. For him, manipulation is an inevitability of design and an inevitable aspect of a free society:

'Yeah, I think there are a lot of discussions that we should be having about how the knowledge that we gain from this enterprise might be used because [...] there are ways in which it may actually be used, not necessarily in a positive way, like the fact that if I go into a shopping mall they're manipulating me. I mean, it's a fact of life, but whether it bothers me that it's done? It's unsurprising, and I don't think there's any way in free society that we could ever control that.' (interview, June 2013)

Neuroarchitecture faces three distinct difficulties here. First is the issue of how the 'environment' or context that shapes human activity is conceptualised in the first place. Is it adequate to maintain that human activity, let alone behaviour, is shaped only by its immediate environs? Does this contention not deny the significant role of historical power relations in shaping the built environment, the role of global and local drivers in shaping interconnected places and the non-proximal, mediated nature of much human interaction or behaviour? Second is the intractable problem of bringing lab-based insights on the ideal conditions for visual perception into the complex social world in which the ideals to be pursued are necessarily a question of political argumentation and struggle rather than scientific forms of reasoning. Thus, to add neuroscience to the E–B framework of environmental psychology seems to go against one of the core purposes of this latter field, which was to understand the mind and behaviour *in situ*. And finally, the interdisciplinary enterprise of neuroarchitecture must contend with an observed reluctance of neuroscientists to seriously engage with architects on an equivalent level. As Tom Albright recounted, there are neuroscientists who want to teach architects about visual perception. He himself feels he has been influenced by architects in terms of how he designs physiological experiments, but he also expresses concern that ANFA has not reached a sufficient number of neuroscientists, and that there is much scepticism among the neuroscientific community about the project of bringing architecture together with neuroscience in this way. In the following example I examine these three potential sticking points in relation to designing health spaces.

Designing health(y) spaces

The second example of neuroarchitectural work that I want to explore relates to the design of health, care, therapeutic and adapted spaces that are shaped around the needs of people who are sick, elderly or living with disabilities. As noted above, the design of healthcare and therapeutic places is an important part of the emerging repertoire of neuroarchitecture, with Ulrich's (1984) study of the positive impact of windows with views of nature on the recovery of patients from (gall bladder) surgery proving an important reference point for contemporary contributors to neuroarchitecture (as cited in Sternberg, 2009; Barrett et al, 2013). Contemporary environmental psychologists have also taken a close interest in this kind of work, in the search to accord 'nature' a psychological or therapeutic value (Steg et al, 2012). More specifically, the E/B/N framework has been developed in relation to observational behavioural research in Alzheimer's care spaces, in which the connections between memory, place and wayfinding (including reducing visual 'clutter') are considered crucial to the functionality of the environment and the experiences of older people living with dementia (Zeisel et al, 2003; Zeisel, 2006). Perhaps the most well-known contributor to this field of enquiry is Esther Sternberg, Professor of Medicine and director of the University of Arizona Institute on Place and Wellbeing, and former ANFA board member.

One of her popular titles, *Healing spaces: The science of place and well-being* (Sternberg, 2009), has explored how both physical and spiritual spaces can have a powerful effect on recovery from illness, through her investigations of the interrelationships between the senses, the emotions, the immune system and place. Such was the book's popularity that she developed and chaired a television programme of the same name in the US. Her contributions to understanding spirituality and healing have seen her invited to meet both the Dalai Lama and Pope Benedict XVI.[3] Credited with several important scientific discoveries across her career, including understanding that the feeling of stress can cause physical ill health, Sternberg now advises several institutions on the relationship between the natural and physical environment and health, including the US Green Building Council, the American Institute of Architects (AIA), the US General Services Administration, the Institute of Medicine and the US Department of Defense. Recalling her experience of an AIA workshop at Woods Hole, Massachusetts, where the idea of ANFA was first mooted by J.P. Eberhard, Sternberg (2009, p 6) summarises that for the participants, Ulrich's seminal study on the value of hospital spaces with windows overlooking natural environments was taken as read. The real question was now to understand '*how* the healing mechanism operated. What brain pathways did windows and their views of nature activate? And how might these affect the immune system and its healing process?'

For Sternberg, the coming together of neuroscience and architecture holds much promise to 'understand how physical surroundings affect emotions and how emotional responses to architecture affect health' (2009, p 7). This would allow for the scientific design of buildings that could contribute to people's health in much the same way that modernist architects such as Frank Lloyd Wright, Richard Neutra and Alvar Aalto had envisaged (Sternberg, 2009, p 5). But her account is also suggestive of some of the potential barriers in developing an interdisciplinary neuroscience and architecture paradigm. In isolating the effects of windows on post-surgery healing, Ulrich's original research had to 'control for' several personal factors (sex, age, past medical history, whether patients were smokers) and other characteristics of the room, leaving him with only 46 participants. In this way, the 'environment' of the hospital to be studied (as in the aforementioned HEAD schools project) is necessarily reduced to the immediate surroundings, and within that, to the window itself. In isolating this factor as the key variable, the capacity of such research to understand the broader environmental determinants of health and healing has to be diminished. The compositional factors relating to who the patient is (class, age, sex, race, background, medical history and so on), and potential contextual variables that may affect healing, are factored out in this model. These may relate to such varied phenomena as how many visitors the patient has, how well supported they feel,

---

[3]   http://integrativemedicine.arizona.edu/about/directors/sternberg

their past experiences of ill health, the care they receive from hospital staff, the employment terms and conditions of those staff, and the way in which medical care is paid for. Such reductionism may be entirely necessary as part of the scientific method pursued here, yet it also indicates some of the limitations of this approach to understanding the relationship between people and their real-world environments.

Sternberg is well aware of these potential blind-spots, and indicates that overcoming disciplinary divides is the only way to address them. Indeed, she states that the inclusion of environmental psychologists in the AIA conference, often derided by architects and neuroscientists alike as a 'soft science' based on mere observations and self-report questionnaires, indicated something of a rapprochement in this regard (Sternberg, 2009, p 8). In asking how windows affect healing, she explains that architects are well-placed to measure physical space variables such as 'light intensity, wavelength, and color; temperature; airflow; and levels of activity in the scene being viewed' (Sternberg, 2009, p 9). Neuroscientists, for their part, could monitor brain activity, physiological response, stress hormones, heart rate, breathing, immune response, drug doses and so on. Yet all these variables, she astutely argues, may still miss out the many intangible aspects of places that might make it good for healing, such as whether 'the most important thing a window does is provide a portal – an escape from the frightening, painful reality of disease, or a way of accessing memories of a better time and place' (Sternberg, 2009, p 9). It could be sensibly speculated that this more personal and sociological role of the window cannot be fully understood by regard to either the physical properties of the space or the biophysical properties of the body–mind nexus. Dialogue between neuroscience and architecture, even tempered by environmental psychology, may not therefore provide the fullest possible explanation of the contextual complexity of healing places.

In a similar vein, Margaret Tarampi's work has focused on the potential of neuroarchitecture to improve spaces for people with particular health needs. She has also explored the characteristics and training of people who are regarded as spatial experts, such as architects or dancers, so that lessons may be learned from their 'spatial ability' for others who are less 'spatially able'. Her research to date has considered how best to design spaces with people with visual impairments in mind – specifically those with low (rather than no) vision who are not currently covered by statutory building regulations under the American Disabilities Act 1990. Her challenge, working alongside computational neuroscientists, computer scientists, theatrical lighting designers and other behavioural psychologists (a cadre of experts redolent of Thrift's

[2004, p 68] account of affect engineers), is to develop software that can be added to computer modelling programs used in architectural practice which would "predict – based on the geometry, the lighting, the materials of a space – if something may be problematic for someone who has low vision problems" (interview, June 2013). She considers this a highly complex challenge in terms of the algorithms (the significance of which is further explored below) required for such software. This is because light, for instance, changes throughout the day and the year. These changes have rarely been taken into account in architectural practice, but have potentially huge ramifications for the experience of a space for someone with low vision. The ultimate aim is to change the American Disabilities Act to include regulations that would address the architectural needs of those many people living with low vision.

Margaret Tarampi is no straightforward 'neuroarchitect', if any such profession could yet be said to exist. Indeed, she admits to having been slightly frustrated by the inadequacy of neuroscience to address the more applied questions around how people actually function in space with which she has been concerned:

> '... [neuroscience is] asking these fundamental questions about the nature of, you know, how the brain works and how that's related to architecture maybe or just how the brain works in general. And that, depending on the question, may be pretty far removed from anything that would then result in changes in how architecture is practiced.' (Maragaret Tarampi, interview, June 2013)

Rather, despite her professional experience as part of a small and select group of scientists and architects developing the cadre of 'neuroarchitecture', Tarampi, too, remains somewhat sceptical about the promise of the neurosciences in general to satisfy the kind of questions she and others at ANFA have been asking, questions which are more behavioural and less reductionist than those concerned with neural processes per se:

> 'Well I think psychology is reductionist as well [as neuroscience], I mean, just the scientific method is inherently that way, but I mean, I think that's part of what draws me to psychology in terms of being able to study some more complex phenomenon that you can't necessarily study maybe in neuroscience. But I would say my experience in terms of talking with neuroscientists who've been involved

in this neuroarchitecture, I think conceptually they're very interested in the idea, but I think they may, I think a lot of them do share my point of view, but that, you know, neuroscience maybe, there's a lot of work that needs to happen in between neuroscience and some sort of applied work in architecture.' (interview, June 2013)

This again points to some of the difficulties that have been experienced in trying to forge new alliances between architects and neuroscientists. It suggests that the findings of neuroscience are still not easily applicable in real-world settings and within pre-existing regulatory infrastructures. Margaret Tarampi speculates that the kind of complex computer algorithms necessary to make practical sense of research on the neuroscience of vision may take several years to develop. Furthermore, as those who may have been involved in disability rights activism will already know, the journey from understanding the needs of people with vision impairments to addressing those needs is a drawn-out one which is political in nature. There is a lot of 'noise', therefore, between the findings of brain scientists and the endeavours of architects with which this alliance must contend. The troublesome issues of defining environments beyond immediate surroundings, translating neuroscientific research into physical spaces, and forging research dialogue between neuroscience and architecture seem to recur across the examples of neuroarchitectural research and practice recounted to this point. But nowhere are these issues more prevalent than in attempts to design and deliver behavioural environments within the sphere of urban design, planning and governance, as the following section considers.

Urban design of behavioural environments

For some working at the intersections of neuroscience and architecture, such as Margaret Tarampi, the neuroscientific examination of a person's spatial perception and cognition, and their experience in the wider urban environment, are not directly comparable. Tarampi's work concerns personal embodied space (within arm's reach) and vista space (in view, but out of arm's reach), rather than the neuromolecular functions of the brain or the wider geographies of the city. While she considers that environmental psychology may well shed light on how we navigate through a building or even a city, she also maintains that the emerging findings of neuroarchitecture are not scalable (Maragaret Tarampi, interview, June 2013). However, others, such as Eve Edelstein, another ANFA research fellow and neuroscience, anthropology and architecture expert at Sternberg's Institute of Place and Wellbeing, University of Arizona, are more buoyed about the potential for '[t]ranslating neuro-architecture from cell to city' (Edelstein, 2014). This extension of the scale of inquiry from neural processes to urban experience has involved methodological and technical innovations in the shape of immersive computer-assisted virtual environment (CAVE) visualisation technologies developed by the University of California Institute of Telecommunications and Information Technology (CalIT²) in the late 2000s. The CAVE is a four-dimensional virtual reality space, a temporary 'room' within which research subjects can move through simulated buildings or cities. Edelstein has been part of a research team that has used the CAVE to study subjects' brain activity using EEG, to record their visual attention using an eye-tracking device, as well as other psychological and biophysical dynamics such as heart rate and galvanised skin responses to measure arousal, attention and memory. The CAVE apparatus, along with new techniques in 'sound bending' (which can direct specific noises, that is, speech, and dull out interfering noise), and advancing evidence on the effect of light on the human body clock and sleep rhythms are informing the design of health spaces (Edelstein's own work on a hospital campus in China and a medical and mental health facility in Canada). The revolutionary potential of these new technologies relates to their capacity to record brain (and bodily) response in mobile subjects, overcoming one of the major shortcomings of fMRI scanning of horizontal subjects in loud, immobilising machinery in the neuroscience lab (Edelstein et al, 2008). This type of measurement of biophysical and psychological parameters has become more commonplace outside of the discipline of psychology, with human geographers, architects and urban planners beginning to use skin galvanic measures, heart rates and geo-located psychometric tests as evidence of research subjects' experiences of space and levels of subjective wellbeing or stress (see, for example, Nold, 2009; J. Anderson, 2012; MacKerron and Mourato, 2013; Resch, 2013; da Silva et al, 2014).

Yet the research enabled by CAVE technology remains virtual in nature – whether three-dimensional or four-dimensional, allowing movement and change over time, it is still only a simulation of reality. Conversely, measuring phenomena such as stress through skin conductance in non-experimental conditions outside of a laboratory carries with it a number of potential methodological limitations, such as how to identify and validate the independent causes of stress in any given situation. These methods tend towards a rather limited conception of situations, as explored in Chapter Two. But arguably these limitations should not be the main concerns for social scientists concerned with understanding the implications of the neurosciences for policy and practice.

This section sets out three interrelated concerns related to the use of both neuroscientific and environmental psychological research in urban design, and outlines some of their potential ramifications for changing governance and citizenship in the context of city life. First, problems are identified in terms of scaling up the *neuromolecular gaze* in the design of 'healthy cities'. Second, the problem of expertise is explored in relation to who should have the remit to change people's behaviour through spatial design, examining how the citizen is urged to both *govern the self* and *be governed through the brain*. Finally, problems relating to the political economies of attention are highlighted in terms of how specific interests might be served in directing people's attention in these ways. How might this serve to *normalise* and *subjectivise* the citizen of the neuroarchitectural city? The aim – as throughout the book – is not to be alarmist here about the potential for neuroarchitecture to be manipulating. Since all design is aimed at influencing its target audiences and users, such a critique would be obsolete. Instead, the aim is to sound a note of caution regarding the use of neuroscience and environmental psychology in urban design, to identify the role of such knowledges in the making up of citizen subjectivities, and to retain a healthy scepticism towards truth claims relating to situated human behaviour. These cautionary notes support an analytical space to open up what counts for evidence in the explanation of the character of human experience.

Such problems or questions, it should be noted, have a long history. Margo Huxley (2006), who we encountered in the previous chapter, helpfully provides an historical overview of the spatial rationalities that have shaped the contexts in which people are governed through urban design. Deploying a Foucauldian theoretical frame, she draws attention to both the importance given to governing fleshy and vital bodies, and the discursive ideals of city planners. Her first account is of a *dispositional* spatial rationality, exemplified by the precursor to Ebeneezer Howard's 'Garden City' and the 'Model Town' movement in the mid-19th century, in which urban design was aimed at overcoming disorder. This was aimed at ensuring healthful living and securing social harmony. Organised sanitation, access to open spaces, light and ventilation, the provision of public spaces, having a clear hierarchy of dwelling types, facilitating visibility and surveillance of inhabitants by one another, these were all part of the urban design strategies deployed in these cases. The second spatial rationality overlaps with the first somewhat, and is *generative* in that it draws on medical and biological theories in order to justify the shaping of environments in order to avoid moral and physical 'disease'. This reformist city of the 1870s-1980s became analogous to the body, and all impediments to circulation were to be removed – blockages such as concentrations of unhealthy, poor or otherwise morally 'corrupted' inhabitants were to be cleared or dispersed, and the environmental context itself was held up as a cause of and potential solution to lack of moral and physical health. Sanitation, air quality and circulation of goods and people were the main means by which the hygienic bio-social city was to be achieved (Huxley, 2006, p 780). By the end of the 19th century, Huxley argues that the dispositional and generative regimes were supported by a *vitalist* spatial rationality that drew on philosophies of vitalism, spiritualist ideas, and the notion that humans had a synergistic relation to their environments.

Urban environments were understood to shape not only human health and morality, but that the minds and spirits of their inhabitants adapted in response to the experience, engagement and memories of cities. This was a 'progressive' rationality in the sense that the city was seen to have a civilising, creative and evolutionary effect on its citizens. In the following sections, I outline how neuroarchitecture, where it can be used in the governing of urban life, has the potential to pre-figure a new *brain-based spatial rationality* which: (1) aggregates neuroscientific data into regimes of algorithmic city governance; (2) privileges the (often privatised) expertise of neuroarchitects and behavioural architects in shaping spaces (as environmental *situations* rather than geo-historical *contexts*); and (3) transforms urban life into a carefully pre-scripted experience through the governance of attention.

Scale: from the neuromolecular to big data

As already noted, the use of *in situ* or geo-located biophysical measures or simulated virtual reality environments indicates a recognised need to overcome the reductionist focus of the *neuromolecular gaze*, which, as Abi-Rached and Rose (2013, p 43) highlight, poses the observable organic functioning of chemical and electrical transmission in the brain as the definitive explanation for human behaviour. The framing of scale has implications for the production of power relations in urban governance (McCann, 2003). Contemporary policy initiatives for designing and creating healthy, sustainable and liveable cities involve a combination of a neuromolecular *and* molar scale, the latter connoting attention to the more readily observable (and often therefore behavioural) parameters of the human body (Kraftl, 2014). The focus on behavioural forms of governance of this sort, where solutions are posed at the scale of individual bodies and behaviours, arguably carries with it a blinkered view of the causes of human action or inaction. Claire Herrick (2009), for instance, argues that the UK's various policy initiatives aimed at creating 'healthier' or 'fitter cities' which encourage healthy eating and physical activity are limited by 'toolkit thinking' through which urban designers tend to regard human behaviour as directly determined by physical surroundings at the micro scale. The perhaps obvious (but no less significant) criticism of this 'environmental input–cognition–behavioural output' framework indicates some of the problems associated with shoe-horning human behaviour into a scientific model which may be valid within a laboratory setting, but which, when it promises to predict action *in situ*, may fail to deliver substantive health outcomes. Whether a particular place promotes and supports behaviours such as cycling and walking to work, for example, arguably has little influence over fundamental spatial inequalities in health and disease – significant barriers to 'scaling up' local policy successes remain.

Similar criticisms can be levelled at Thaler and Sunstein's (2008) notion of 'choice architecture' where they propose using spatial prompts to work around human biases in decision-making and apparently self-defeating judgements. Yet it would be churlish to regard the policy imperative for healthy cities or the pursuit of places to promote wellbeing as undesirable goals. Clearly government action to improve people's experiences of places, and to facilitate their flourishing as individuals and communities, is commendable. What may be at issue, however, is the way in which 'wellbeing' and 'place' are conceptualised within policies aimed at designing healthy, sustainable and liveable cities. As Atkinson et al (2012, p 4) have argued, such policies are reliant on a specific operationalising of the concept of wellbeing. This is possible first, by breaking wellbeing down into its constituent components so that it can be more easily measured. Second, the dominant policy understanding of wellbeing relates to subjective or objective experience at the level of the individual, with contextual factors (for example, the economic wellbeing of a nation, community notions of the good, collective commitments to social and environmental justice) regarded as somewhat marginal to the aggregate of individual wellbeing (Atkinson et al, 2012, p 5). Third, wellbeing has often been conflated with health, specifically mental health (or the absence of mental illness), and has become synonymous with the vague notion of psychological resilience (Atkinson et al, 2012). Atkinson has therefore cautioned public health practitioners and policy-makers against a narrow scalar focus on the individual experience of subjective wellbeing at the expense of 'understanding the social, economic and environmental factors within which the need for resilience emerges in the first place' (Atkinson, 2011, p 2).

The idea of urban liveability and place-based wellbeing are fundamental to the phenomena of the 'Smart City'. The Smart City is also arguably one way in which it is sometimes hoped that the principles of neuroarchitecture can be scaled up in changing and governing behaviour through urban design and management – through the aggregation of individual neuroscientific data into population-level databases of biophysical information. The notion of the Smart City denotes an approach to urban planning and governance which is committed to sustainable economic growth, improvements in quality of life, high levels of social capital and the existence of a sophisticated digital infrastructure that can be used in 'enhancing the competitive profile of a city' (Caragliu et al, 2009). Its existence relies on the aggregation of individuated data that is usually geo-located and can be temporally specific and/or in real-time; so-called 'big data' (vast datasets that are difficult to organise, store and analyse). Critics have already started to highlight the potential ethical challenges posed by the advent of big data in the neurosciences, particularly surrounding the US BRAIN research initiative. Choudhury et al (2014) point to: the divergent research and political goals of the project's public and private sponsors; the need for measures to protect individual privacy and potential harm to specific social groups of scientific generalisations; and the challenges posed for openness, transparency and data sharing. In the case of Smart Cities, there is an increasingly large amount of data recorded relating to how people, materials and information move around urban space. Here, it is often associated with multinational technology company IBM's invocation of the Smart City as a solution to tightened state budgets,[4] and includes IBM's own proprietary software designed for city leaders to manage resources, gather intelligence and make strategic decisions.[5]

The oft-heralded dawn of big data promises to revolutionise how urban spaces are governed and managed. Databases may include information from social media, from sensors, GPS, mobile phones

---

[4]   www.ibm.com/smarterplanet/us/en/smarter_cities/overview/

[5]   IBM® Intelligent Operations Center, www-03.ibm.com/software/products/en/intelligent-operations-center

and internet use, and such data may be owned by corporations or governments. There is also potential for biometric data to be collated in this way that would clearly lend itself to use by neuroarchitects. The potential aggregation of neuroscientific and behavioural data in this way would render geo-computation and urban informatics experts, technology companies and the holders of such datasets extremely powerful. Critics have already highlighted how the Smart City could be used as 'a means to discipline cities and their populations, reducing sustainability and the urban question to technical discourse' (Gibbs et al, 2013), or as merely 'a high-tech variation of the "entrepreneurial city"' (Hollands, 2008), where cities must compete with each other for the attentions of investment capital. The advent of logistical management of biological data implied by a potentially scaled-up neuroarchitectural infrastructure must therefore be accompanied by modes of social and cultural analysis which can decipher the contextual rationalities that are materialised by this arrangement.

Others have gone further to suggest that the big data revolution has led us into an age of 'algorithmic governance' (Williamson, 2014), whereby computer code, algorithms and software are mobilised in the pursuit of public policy goals and as a means of knowing and governing the current and future behaviour of citizens. In this way, the algorithm itself has power, which 'functions through collecting, collating and calculating the data of citizens in order to predict their probable future needs, and by automating the process of personalisation' (Williamson, 2014, p 310). Williamson (2014, p 297) demonstrates how public policy discourse has shifted towards a cross-sectoral set of networked institutions committed to 'postliberal' forms of governance enabled by a celebratory language of a new 'social media science', whereby vast amounts of personalised data can be used in the proposition of new governance solutions. Similar concerns have also been expressed about software and data relating to the 'programmable city' in relation to how they both translate the city into code and transduce that code into a reformulated urban life: '[software] codifies the world into rules, routines, algorithms, and databases, and then uses these to do work in the world to render aspects of everyday life programmable' (Kitchin, 2011, p 945). Kitchin argues that code, far from being neutral, is 'the manifestation of a system of thought' (Kitchin, 2011, p 497) which is 'shaped by the abilities and worldviews of programmers and system designers working in companies situated in social, political, and economic contexts.' Furthermore, the production of software, Kitchin contends, has its own political economic context, its own discursive registers and its own vested interests that must be examined in relation to the kinds of policy developments for which they have been mobilised. This code space has real effects. As such, where big data and the Smart City have the opportunity to incorporate behavioural, sensory, biophysical and neuroscientific information in this scaled-up manner, we must carefully examine and understand their use in governing and disciplining urban living – just as scholars have interrogated the 19th-century use of social statistics in strategic urban governance (Osborne and Rose, 1999). That the experts of neuroarchitecture and the programmers and system designers of the Smart City are all geared towards *securing optimised behaviour in situ* implicates urban architectures in the governance of the future of life itself (Kraftl, 2014, p 9).

Expertise: corporate and public neuroarchitects

There is much at stake in claiming the authority to explain the spatial determinants of human behaviour, and in assuming the mandate for shaping those behaviours through architecture and spatial design. Furthermore, the scale at which such explanation is offered – from the neuromolecular scale of the neuroarchitect through the molar scale of the environmental psychologist to the often multi-scalar analyses of the social scientist – has broad ramifications for the kinds of policy solutions offered for problems that are framed as behavioural and/or biological. As such, a prominent criticism of the 'choice architectures' approach of nudge-inspired policies in the UK is that such behavioural 'prompts' are too small-scale, too trivial to be used to govern whole populations (House of Lords, 2011). Critics have also noted that the imperative to govern behaviour through spatial design, whether focused on 'situational' crime prevention, urban security, sustainable travel, obesogenic environments or liveable neighbourhoods, relies on an impoverished conceptualisation of space which fails to appreciate the many scales at which human practice operates (Jones et al, 2013, p 108). It is something of a conceptual leap to extrapolate out from neuroscientific studies of our sensory perceptions and embodied cognition of our immediate physical surroundings towards the prediction and governance of our behaviours in specific social, cultural, economic, political, historical and geographical contexts. And yet there is much contemporary architectural and design research and practice committed to using molecular- and molar-scale behavioural scientific knowledge to address some of today's biggest global challenges. One such initiative, a partnership between the Design Council and behavioural scientists at Warwick Business School (WBS), the

Behavioural Design Lab, posits the psychology of human judgement and decision-making as the driving force behind obesity and climate change (Gardiner, 2012, p 1), proposing design solutions based on randomised behavioural science experiments.

Moreover, the very definition of architecture and design are being challenged by the popularity of behavioural economic and neuroscientific approaches to understanding decision-making, with market researchers and communications companies now advertising themselves as architects of behavioural change (for example, the global market research consultancies, The Behavioural Architects,[6] and UK-based Ogilvy Change[7] and Corporate Culture[8] and, formerly of the UK's Cabinet Office, The Behavioural Insights Team being just four examples of a growing international behaviour change industry). These largely private consultancies are complemented by an increasing number of public sector institutions, civil servants, publically funded researchers and public–private bodies seeking architectural solutions to what are deemed to be behavioural problems. Public health researchers, for example, have become interested in public health interventions around smoking, diet, physical activity and drinking 'that involve altering small-scale physical and social environments, or micro-environments' (Hollands et al, 2013, p 2). This can literally involve making high-calorie foods 'slightly more difficult to reach' by only 10 inches in a cafeteria salad bar (Rozin et al, 2011), scientifically validating one of the introductory petitions of Nudge to 'place the fruit at eye-level' in order to combat the global obesity epidemic. In this example, the molar-scale of our embodied proximity to healthy food choices, and distance from unhealthy ones, is posed as the solution to the public health problem of obesity. We are impelled to *govern the self* by resisting temptation, but we are enabled to do so by the psychological targeting of the automatic mind, as set out explicitly by public health researchers Marteau et al (2012). In other words, we are *governed through our brains* by experts in brain- and body-based explanations of our behaviour. Who these experts are accountable too, the methods and principles they prioritise, and the purposes to which they put their expertise are thus important questions relating to the democratic legitimacy of neuroarchitecture and urban design.

---

[6] www.thebearchitects.com/home

[7] www.ogilvychange.com/

[8] www.corporateculture.co.uk/

Attention: from eye-tracking to urban experience

As many of the aforementioned examples demonstrate, neuroarchitects and behavioural experts alike have expanded our understandings of how humans make decisions. They have marked a radical departure from classical economic accounts of 'rational economic man', and have supplanted him with a far more nuanced and experimentally validated set of explanations for behaviour *in situ*. These explanations have been based on new technological developments which allow scientists to investigate the relationships between reflexive and automatic brain processes, between the rational and emotional, between biophysical, affective responses to the world and expressed feeling, between the body and the mind. Indeed, some of these dualisms have been effectively shattered by neuroscientists and environmental psychologists. There is great promise that these new forms of behavioural evidence will provide the basis to make better educational and health spaces and more sustainable, healthy and liveable cities, to fulfil many lofty ideals. And there is also now the means to incorporate such data in the pursuit of personalised forms of algorithmic governance in the programmable city. The findings of these new experts in behaviour therefore carry with them a new set of concerns and a new set of responsibilities if they are to avoid promoting a blinkered view of the causes and consequences of spatial practice which limits itself to a diminished understanding of the immediate and proximate situations in which people make decisions. If the potential biological determinism of brain-based explanations or the environmental determinism of architectural and design propositions are to be avoided, we must persist in expanding our notions of what drives behaviour, where decisions come from, and how human subjectivities are shaped. We should not be satisfied with accounts that reduce the geo-historical context to local surroundings/situation, and we must continue to problematise what counts as evidence within the fields of urban policy and practice.

Paying attention to the way in which our attention is always already mediated by our geo-historical context can go some way to achieving these modest goals. In this sense, our conscious or subconscious experiences of urban life cannot be fully understood by measuring our brain activity, biophysical markers or self-report questionnaires. Many aspects of urban experience simply evade measurement or collation as evidence. Take, for instance, German cultural critic Walter Benjamin's notion of the city as a capitalist spectacle, Marxist geographer David Harvey's conception of how 'capitalism produces its own geography' (Harvey, 1989, p 5), or Henri Lefebvre's (1991 [1974]) account of how space itself is socially produced. If the materialism of our urban experience is to be understood geographically and historically, there is a need to address the specificity of that urban experience in terms of the political, economic, social and cultural driving forces that influence our behaviours, shape our subjectivities and direct our attentions in particular ways within particular spaces.

Advances in neuroarchitectural technologies and techniques – from the recording of galvanic skin responses and EEG, through eye-tracking and 'sound-bending' hardware, to the potential aggregation of such personalised responses in proprietary Smart City databases – raise difficult questions about who is directing our attention, to what ends, and at the expense of attending to what alternatives. In a critical reading of the political economies of attention, Matt Hannah puts the case that attention 'is a scarce commodity on the way to becoming as important a currency as money in the mediatized world of twenty-first century capitalism' (Hannah, 2013, p 244; see also Crogan and Kinsley, 2012). Whether it is the intense rigour of neuroscientific evidence or the potential value of the Smart City's quantitative databases for knowing human behaviour, there is a significant risk that these narrow forms of evidence occlude more expansive and politicised ways of understanding the character of urban experience. It is these more contextualised understandings – which take into account both the embodied experience of the city and the discursive scripting of that experience – which provokes us to question what is the democratic legitimacy of these new experts in architecture and urban design? Who decides on what social goods will be pursued by new techniques for directing our attention? And in these neuroscientifically, behaviourally and socially idealised spaces, what room is left for attending differently, beyond the formulaic inscription of our mental and embodied responses? In this sense, the constitutive role of research and practice in neuroarchitecture and environmental psychology in potentially *normalising* human responses and *subjectifying* idealised citizens as identified and individualised through big data requires interrogation. The specific ways of knowing human behaviour implicated by neuroarchitecture and environmental psychology are partial truths relating to human conduct and subjective experience. It seems increasingly likely that these partial truths will form the new databases of algorithmic governance through which the attentions and behaviours of citizens can be modified. As such there is a need to carefully examine their use in the governing of urban life.

FOUR

Teaching the learning brain

The way in which brains operate is *fully determined* by the integrative properties of the individual nerve cells and the way in which they are connected. It is the functional architecture, the blueprint of connections and their respective weight, that determines how brains perceive, decide and act. Hence, not only all the rules according to which brains process information but also all the knowledge that the brain possesses reside in its functional architecture. It follows from this that the connectivity patterns of brains contain information and that *any learning, ie the modification of computational programs and of stored knowledge,* must occur through lasting changes of their functional architecture. (Singer, 2008, p 98, emphasis added)

You hold in your hands a historical publication. This book is the first to bring together some of the most influential scholars responsible for giving birth to a new body of knowledge: *educational neuroscience.*[...] And teaching will never be the same again. (Sousa, 2010, p 1, original emphasis)

Introduction

'Neuroeducation', including in its various guises, the research endeavour of educational neuroscience and more applied approaches to brain-based or brain-compatible teaching and learning, seems to signify a more developed and defined field than neuroarchitecture. Somewhat confusingly in light of the previous chapter, it is the neuroarchitecture or functional architecture of the brain (the organisational networks and connections of the brain) that preoccupy neuroeducationalists. There is often a tendency in this work to understand this functional architecture in explicitly determinist terms, and to reduce learning to an algorithmic or computational process, as in the account offered by renowned neuroscientist Wolf Singer (2008) in the above quote. Neuroeducation has been promoted since the 1990s through learned institutions, academic publications, corporate organisations and consultancies, and promises to revolutionise the world of teaching and learning, often in the hyperbolic manner offered above by Sousa. Broadly speaking, neuroeducation aims to use neuroscience, biology and cognitive science to better understand how people learn (usually focusing on children in formal educational settings).

There has been a concerted effort to separate out evidence-based pedagogic techniques and so-called 'neuromyths', and to elaborate the grounds for cultivating two-way dialogue between neuroscientists and educators (Blakemore and Frith, 2005; Goswami, 2006; Howard-Jones, 2007; The Royal Society, 2011b). Indeed, a clear distinction is sometimes drawn between educational neuroscience and neuroeducation, the first signifying the pursuit of empirical scientific research, using brain imaging technologies such as fMRI and EEG, and other neuroscientific methods – to investigate neural processes considered to be most relevant to the learning brain – and the latter indicating the application of neuroscientific findings and theories to classroom practice. This chapter focuses on the latter in considering the governance and citizenship implications of pursuing a neuroscientific approach to teaching and learning. However, in order to fully understand the evolution of neuroeducational practice, it is necessary to elaborate on the relationship between educational neuroscience and neuroeducation, including the efforts that have been made in bridging disciplinary divides and the divide between science and practice. This chapter draws on qualitative interviews conducted between October and December 2013 with educational psychologists (EPs) based in local education authorities in the UK, educational consultants who offer training/services in and writing on neuroeducation, teachers who have enthusiastically embraced these new approaches to the 'learning brain', and members of the Royal Society's Working Group on Neuroscience and Education (The Royal Society, 2011b). Together, they make up a diverse group of interlocutors between neuroscientific research and educational practice.

The chapter begins by outlining some common research issues within the field of neuroeducation before moving on in the second section to explore the key features of debates surrounding its inception as a distinct body of knowledge. This includes the apparently contradictory efforts to both *bridge* and *distinguish between* neuroscientific research and educational practice. While the science of learning is often conceived as its own rationale, the educational rationale (relating to the context-specific social purposes of educational structures and institutions) for pursuing pedagogies based on brain science is less often explored. This second section thus investigates some of the given reasons for a need for neuroeducation. A third section explores the emergence of brain-based teaching practices in the context of economic, educational and technological change. The fourth section examines the particular perceptions of EPs who have been exposed to neuroeducation in their professional training and development in order to better understand some of their conceptual and practical concerns in adopting neuroeducational approaches in contemporary formal settings. Given their work with students already labelled with learning difficulties, emotional and behavioural problems and psychological disorders, EPs are well placed to consider the specific role of neuroeducation in the governing of psychological subjectivities and shaping behavioural norms. The final section further extends these considerations to introduce some of the wider criticisms levelled at neuroeducation from sociological and philosophical perspectives. It examines the potential implications of an emerging brain culture within the formal school system for learning, and of developing a new cultural relationship with our brains.

Research topics in the neuroscience of learning

There is a wealth of neuroscientific research on learning, addressing the processes by which learning takes place in the brain, the biological and environmental drivers of brain development, and learning difficulties and disabilities (predominantly dyslexia, dyscalculia, autism, Asperger's and ADHD), which are summarised briefly here in six areas of research and practice. First, much of the existing neuroscience research on learning relates to learning in the particular sense of numeracy and literacy skills (investigated through specific tasks set by researchers), although there is also recent research on creativity (Benedek et al, 2014) and the development of social, moral and emotional understanding (Blakemore and Choudhury, 2006). A second set of research issues relates to various aspects of the embodied brain and how best to care for it in order to optimise learning. Studies reported in the Economic and Social Research Council's (ESRC) Teaching and Learning Research Programme (TLRP) on Neuroscience and Education (Howard-Jones, 2007) have investigated the role of sleep on memory, the importance of exercise and nutrition for learning, including assessing the detrimental role of caffeine on alertness and mood, a passing fad for encouraging drinking water in classrooms, and controversial studies on the use of fish oils to enhance educational performance in tests.

A third area of research which has been the subject of much recent policy debate in the US and UK concerns the existence of critical periods of development (most notably the age range 0-3), the contention that children can only learn specific skills or functions during specific windows of opportunity, and that where the environment in which the child is brought up is educationally limited or impoverished, these windows are lost forever. This was taken by some as evidence of the need for 'early intervention' programmes in the guise of more intensive social work, parenting programmes and pre-school curricula. However, it is now understood that the brain is capable of structural and connective change, often captured by the phrase 'neurons that fire together, wire together'. The plastic nature of the brain thus signifies the wealth of educational possibility throughout the life course for the ongoing learning and adaptation of the brain. So, too, it highlights that the environment in which the brain develops can have a significant impact on the brain's very biology – re-orienting debates around nature *versus* nurture in ways that emphasise the way in which they work together in shaping learning.

A fourth dimension of neuroeducation focuses on cognitive enhancement – whether this refers to the aforementioned measures to be taken to get the brain in a learning-ready state (for example, nutritionally, through sleep or exercise), or indeed the kinds of brain-based teaching and learning interventions that make up a diverse range of approaches from the most morally challenging (for example, the use of pharmaceuticals in test and exam situations), through commercially available brain-training programmes such as 'Brainology', 'Cogmed' or 'Brain Gym', to more commonly used (and often proprietary) brain-based pedagogies and classroom activities (for example, the Kagan method of brain-friendly learning and brain-based thinking tools,[1] Jensen's [2007] 'brain-compatible learning', or Hardiman's [2012] 'brain-targeted teaching' model). It is regarding this dimension of neuroeducation, which has been driven more by teachers and educational consultants than by neuroscientists, that there has been a recognised concern relating to the potential misappropriation of neuroscience, the need for healthy scepticism and scientific education amongst teachers and the debunking of so-called 'neuromyths'.

The fifth way in which neuroscience and educational research and practice have been brought together is through investigations that address the particular learning needs of specific groups who

---

[1]   www.kaganonline.com/

have been diagnosed with developmental disorders such as dyslexia and ADHD. Neuroscientists have, for instance, examined the neural activity associated with language acquisition, sound recognition and reading skills, and have identified educational interventions which can address and improve difficulties experienced by dyslexic children and adults (Goswami, cited in Howard-Jones, 2007, p 12) – further troubling any supposed division between biological and environmental influences on learning. Identification of the neural 'markers' for these hitherto attainment- or test-based diagnoses is one of the principal purposes of this kind of neuroscientific research, although this is notably in its infancy (The Royal Society, 2011b, p 12). The diagnosis and reported over-diagnosis of ADHD is contested because it often leads to pharmaceutical treatments rather than carefully designed teaching and learning interventions or personalised cognitive training. Despite public controversy concerning the use of drugs such as Ritalin to treat children and adolescents presenting with 'abnormal' behaviours, specifically a lack of 'impulse control' at school, neuroeducators are hopeful that neuroscience and genetics will make diagnosis more accurate (The Royal Society, 2011b, p 13). However, the emergence of a relatively new developmental disorder, known as 'Sluggish Cognitive Tempo' (SCT), is likely only to add to public dissent from such diagnoses. While SCT has not yet been officially recognised as a disorder by the American Psychological Associations *Diagnostic and statistical manual of mental disorders* (fifth edition) (DSM-V), psychologists have argued that it should be distinguished from ADHD (Barkley, 2014). They point to diagnostic ratings provided by teachers and parents through which children are said to show symptoms of daydreaming, drowsiness, slow thinking and absent-mindedness. These children are judged to show many of the symptoms of ADHD without the hyperactivity, yet are still not achieving at school.

For some, these debates should focus on the way in which the structures and environments of schooling funnel children of diverse abilities, temperaments and backgrounds into a relatively narrow set of expectations surrounding their learning-readiness and educational 'performance' (Timimi and Leo, 2009, p 8; Cohen, 2006, p 19). Many parents and advocates, however, strongly believe that an ADHD diagnosis is the first step in a process of securing more educational support for their children, and reportedly prefer this recognition than the social stigma of a badly behaved child (Graham, 2010, p 15). But for many critics it is a categorical error to 'medicalise' these personality traits, particularly in the absence of definitive and objective scientific measures (rather, the diagnostic tests are predominantly reliant on subjective parent and teacher reports). Moreover, the treatment of ADHD with psychoactive drugs poses challenging ethical concerns, particularly given the detrimental effects of labelling children in this way, and a lack of evidence of whether medical treatment actually improves the quality of life and educational outcomes of such children (Smith et al, 2010, p 7). Hence several critics from within educational and disability studies have (for some) controversially argued that ADHD is a socially constructed disorder more driven by social, economic and technological change than by fundamental neural processes (see, for example, Kean, 2009). The geographical and historical differences in diagnostic rates for ADHD in different nation states is evidence enough for many that ADHD is at the very least an ambiguous disorder, the prevalence of which relies on a country's political and cultural approach to education and disability (Jahnukainen, 2010, p 63).

These debates point towards some of the important moral and social issues in scientific research and practice concerning the brain, and signify the high stakes in competing conceptions of the relative importance of the cognising mind, the biological brain and observable behaviour in specific environments (notwithstanding efforts to develop consistent analytical frameworks for the interrelations between them; see Morton and Frith, 1995). The accepted norms and understandings of the causes of developmental disorders also have significant implications for our conceptions of moral and social responsibility, and the legitimacy or otherwise of particular kinds of interventions. While neuroeducationalists may hope that the identification of neural markers and biological processes might take the associated sense of 'blame' away from affected individuals and their families, critics believe that the search for neuroscientific explanations only serves to absolve communities, society and governing authorities of the responsibility to address fundamental flaws in the education system and the more contextual aspects of educational disadvantage. There is often a fine line therefore being negotiated between the diagnosis of developmental disorder and the attribution of more ambiguous emotional and behavioural difficulties among school children and young people.

Emotional and behavioural issues make up the sixth aspect of neuroeducational research that is explored here in what has been a brief and partial introduction to an expansive field. Research on inhibition, self-control and anti-social behaviour, particularly among adolescents, has been prominent not only in neuroeducation, but also in behavioural economic research which has been influential in shaping nudge-type policies in the UK and US in recent years (Dolan et al, 2010). This has been significant in terms of explaining not least impulsive behaviours in adolescents, deficiencies in reasoning and the varying ability of people to resist temptation or delay gratification (The Royal Society, 2011b, p 8). Research on neural development has, for instance, shown that at least two significant developmental changes occur after puberty. These are 'myelination' of the axons (the build-up of a fatty substance which improves the efficiency of neural communication), and synaptic changes ('pruning', whereby new neural connections are made and others lost) in the pre-frontal cortex, the part of the brain associated with executive functions and reasoning, as well as movement (Blakemore and Choudhury, 2006).

Again, these findings have potentially significant implications for our social conceptions of moral responsibility, and such insights have been noted by a group of neuroscientists advocating changes in the legal age of criminal responsibility. On a more everyday level, the apparently innate capacity of the brain to 'self-regulate' (The Royal Society, 2011b, p 8) is leading to educational programmes that aim to support children in managing their impulses, changing ingrained habits and controlling their behaviour. Many of the aspects of the Social and Emotional Aspects of Learning (SEAL) programme introduced in 1997 to promote emotional intelligence in schools focus precisely on this kind of emotional management. As Elizabeth Gagen (2013) has outlined, this cultivates a new form of citizenship informed by new popular neuroscientific conceptions of the emotions. This chapter goes on to explore this new relationship with our brains, and the forms of citizenship and psychological governance invoked by a marked enthusiasm for neuroscientific knowledge within the educational sphere. The following section charts the emergence of a 'learning brain' culture since the late 1990s. It draws out some of the efforts that have gone into creating dialogue between neuroscience and education as distinct disciplines and distinct fields of practice, and considers some of the stated or implied rationales for a perceived need for neuroeducation.

Bridging the divide in the emerging neuroeducation

Despite a considerable amount of neuroscientific knowledge of relevance to learning, neuroscientists and education researchers alike began to observe in the mid-2000s that there had been limited dialogue bringing together research scientists and educators. Distinguished neuroscientists Sarah-Jayne Blakemore and Uta Frith (2005, p 1) outlined the potential for understanding how the brain learns to 'transform educational strategies and enable us to design programmes that optimise learning for people of all ages and all needs.' They describe how dialogue between neuroscientists and educationalists began to take place in the UK with specific regard to a parliamentary subcommittee on early years education in 2000. This debate had been informed by public controversy over the 0–3 campaigns in the US. These campaigns, and the social policies that followed, have been heavily criticised by John Bruer in 1997, who questioned their neuroscientific evidence base, regarding the link between developmental neuroscience and educational policy as 'a bridge too far' (Bruer, 1997).

While the policy reports had drawn parallels with social types of deprivation and the developing brain, the only research that had been carried out to that point seemed to relate to the quite specific scenario of visual deprivation in animals (Bruer, 1997, 2013). Later that year, Blakemore and Frith organised a workshop of scientists, teachers and educational researchers as part of their review of the implications of recent developments in neuroscience for research on teaching and learning for the ESRC's TLRP. They set out specifically to address what they saw as a clear gap between what neuroscientists know about the 'nature of learning' and educational practice (Blakemore and Frith, 2000, p 4). Thus, while they are keen to set out the grounds for a two-way dialogue that avoids the notion of straightforwardly applying brain science to classroom practice (Blakemore and Frith, 2000, p 51), there is clearly a sense in which neuroscientists have something to offer educationalists rather than the other way round. Blakemore and Frith posit cognitive psychology – the theoretical framework for understanding mental processes – as the key mediating discipline between neuroscience and education. This is necessary, it has been argued, because on the one hand, teachers can be too trusting of evidence that appears to be scientific, as with the powerful appeal of neuroimagery in apparently explaining what is going on inside people's brains. But on the other hand, Blakemore and Frith (2000, p 6) also implied that teachers rely too heavily on common sense and professional experience, remaining suspicious of academic psychological research. Writing at the turn of the century, they were nonetheless optimistic that within five to ten years, an interdisciplinary 'learning science' involving neuroscientists, psychologists and educators would have evolved (Blakemore and Frith, 2000, p 51).

Their optimism at this time proved to have some grounds, and the development of neuroeducation has been both rapid and international in scope. In 2003, a group of scientists held a workshop on 'Mind, brain and education' at the Pontifical Academy of Sciences at the Vatican, Rome, featuring scholars from Argentina, Canada, France, Germany, Italy, Japan, the Netherlands, the UK and US (Battro et al,

2008). By 2005, educational neuroscience had started to emerge as a new discipline in the UK, when a new research centre for neuroscience and education was established at the University of Cambridge, with the Institute of Education, University College London (UCL) and Birkbeck following suit in 2008 with their Centre for Educational Neuroscience and a Master's level course. In 2014 another new Master's course was initiated at the University of Bristol, and a joint honours undergraduate degree programme in educational studies and neuroscience was launched at the University of Keele in 2015. The availability of only three such courses across the UK is an indication of the extremely novel status of the field. In the US, a course at Harvard Graduate School of Education on the educated brain began in 2002 (led by Antonio Battro and Kurt Fischer), and later became a Master's course on mind, brain and education, along with a learned society and an academic peer-reviewed journal of the same name launched in 2007. Another international journal, *Trends in Neuroscience and Education*, was to follow in 2012.

Meanwhile, several research publications and editorials have contributed to the evolving field, and much commentary has focused on the possibilities and prospects of an educational neuroscience, from Blakemore and Frith's aforementioned title, *The learning brain* (2005), Goswami's *Cognitive development. The learning brain* (2008), Howard-Jones' work with the ESRC-TLRP (2007), and his book, *Introducing neuroeducational research* (2010), as well as the edited collections, *The educated brain* (Battro et al, 2008), *Mind, brain and education* (Sousa, 2010), *Neuroscience in education* (Della Sala and Anderson, 2012) and *Educational neuroscience* (Mareschal et al, 2013). Critical reviews have sought to distinguish an evidence-based educational neuroscience from its misapplication in brain-based learning packages (Goswami, 2006; Geake, 2008), and have elaborated on the ethical and philosophical challenges of neuroscientifically informed classroom practice (Hardiman et al, 2012; Clark, 2013). A report on the research and policy implications of advances in 'learning science' published by the international policy organisation, the Organisation for Economic Co-operation and Development (OECD) (2007, p 17), noted several other research institutes in countries in which educational neuroscience was gaining traction, including Japan's Science and Technology's Research Institute of Science and Technology for Society, the Transfer Centre for Neuroscience and Learning, Ulm, Germany, and the Learning Lab, Denmark. A fuller history of the emergence of educational neuroscience and national governments' support for education and neuroscience research is provided by Tracey Tokuhama-Espinosa (2011, p 67), whose doctoral thesis at Capella University included a meta-review of the past 30 years of academic literature implicated in the development of educational neuroscience (Tokuhama-Espinosa, 2008).

In sum, these initiatives were united by an enthusiasm for gauging how rapidly advancing neuroscientific and cognitive scientific developments would change how we understand learning, as well as a determination to build bridges with educators in order to shape research priorities and practice in this field. Furthermore, they shared a sense of caution around the premature and unscientific adoption of brain-based teaching and learning strategies. And in distinguishing themselves from the more commercial neuroeducation arena, they have striven to make modest recommendations for educational practice in the limited and specific circumstances where accepted scientific knowledge allows strong conclusions to be drawn, for instance, in relation to the areas of sleep needs, arithmetic, reading and language acquisition (Battro et al, 2008, p xviii). That almost every textbook and research collection on neuroeducation begins with a set of caveats dispelling neuromyths, setting out the challenges of interdisciplinary working, or identifying the incomplete or limited nature of knowledge in this field, is indicative of a sensitivity that the protagonists of neuroeducation have to some of the potential ethical challenges posed by this new discipline. In the coming together of brain science with the everyday education of children and young people, these challenges are brought into sharp focus where commercial companies appeal to the respectability and truth status of science in the promotion of educational products and programmes aimed at teachers and parents operating within national education systems driven by attainment and global competitiveness. Thus, the OECD report, which heralded the birth of a learning science, also warned of the potential impact of the abuse of brain imaging in education, concerns over data protection, cognitive enhancing drugs, the possibility of manipulating behaviour and an '*overly scientific approach to education*' (OECD, 2007, p 17, original emphasis):

> Neurosciences can importantly inform education but if, say, "good" teachers were to be identified by verifying their impact on students' brains, this would be an entirely different scenario. It is one which runs the risk of creating an education system which is excessively scientific and highly conformist.

Yet the endeavour of neuroeducation is self-avowedly driven by the advancement of science; it is posited as a scientifically led discipline. Its science and its devotion to empirical methodologies are what distinguish it from the loose generalisations of brain-based teaching and learning, which should be subject, it is argued, to the same stringent evidential bases and collective scepticism of properly executed neuroscience. The answer, therefore, to the misuse and abuse of neuroscience in education is *more* neuroeducation, the education and training of teachers and children alike in the science of the brain, as one neuroscientist involved in The Royal Society's Brain Waves project outlined:

> 'We were very, very keen to say how important it is at this moment, to educate people about the brain, and about neuroscience and what can be said, and what cannot be said. And you know, to particularly make sure that people like teachers are appropriately sceptical, and that was one of our main concerns, and it is still a concern of mine now. I think this can only really be achieved by giving everyone a proper science education, from early childhood onwards really, to make them savvy and educated in being able to evaluate the claims that are being made in the name of brain science.' (interview, October 2013)

The rise of brain-based teaching: a revolution in practice?

The academic interdisciplinary effort of educational neuroscience, signified by publications, learned organisations and university-based programmes, has been steadily developing since the turn of the century. At the same time, it has been frequently argued that the proliferation of brain-based teaching programmes, initiatives and products has out-paced educational neuroscience by a margin. Yet there has, to my knowledge, been no sociological research on the extent or impact of brain-based teaching in any national context. While there is much anecdotal evidence and media controversy of the use of such techniques (see, for example, Goldacre, 2006; Randerson, 2008), it is difficult to ascertain just how widespread the actual practice of neuroeducation is. A survey carried out with teachers in 2005 found that 78 per cent of teachers and educators thought that an understanding of the brain was important in their educational activities (Pickering and Howard-Jones, 2007, p 110). This research was heavily skewed by the fact that the survey was carried out with teachers attending one of two conferences on education and neuroscience, and an OECD web discussion forum specifically on this topic. Nevertheless, the very popularity of these kinds of events and fora would suggest some degree of enthusiasm among teachers for understanding the learning brain. Furthermore, the survey reported that these teachers were aware of or had been using educational activities linking the brain to education in their schools and colleges. These activities included: teaching and learning approaches (mind-mapping, accelerated learning and brain-based learning); cognitive and neuropsychological approaches based on academic knowledge of these areas; learning styles (including visual, auditory and kinaesthetic [VAK] learning, and left-/right-brain); educational kinesiology (Brain Gym); ingestion and the brain (including fish oils, nutrition and water); and emotional intelligence (Pickering and Howard-Jones, 2007, p 111).

The concern of many educational neuroscientists is that teachers' practice has overtaken science, and that teachers lack sufficient scepticism regarding the brain claims of these educational activities. Yet at the same time as conveying a modest appraisal of the limitations of current educational neuroscience, neuroeducators and educational neuroscientists alike have a tendency to herald a new dawn in education. When asked about the challenges and future possibilities of neuroeducation, a neuroscientist who had been involved in The Royal Society's Brain Waves project responded:

> 'So I think we do need to be aware of how modest our glimpses of brain function are, mostly people don't understand that, mostly people think you can take a photograph of the brain in action and then you can point to a particular spot and say that is where something is wrong – that is just not the case.[...]
>
> 'I am expecting a similar kind of revolution [to that of medical sciences about 100-150 years ago] from what we learn about brain science, and how we can apply that, not just to diseases, brain diseases like you know like mental illness, but to everyday behaviour.' (Interview, October 2013)

The analogy with advances in the medical sciences and the discovery of bacteria, the causes of disease and eventually treatments, vaccinations and antibiotics, is a popular one to be drawn. David Sousa (2010, p 2), a prominent consultant in educational neuroscience and former chemistry teacher based in the US, writes that:

> Teachers have taught for centuries without knowing much, if anything, about how the brain works. That was mainly

because there was little scientific understanding or credible evidence about the biology of the brain. Teaching, like early medicine, was essentially an art form.

His analysis of the transformational potential of neuroeducation is shared by purveyors of brain-based teaching programmes and products, and educators in search of evidence-based approaches to 'what works' in classroom practice.

'100% satisfaction guaranteed'

Sousa recounts the determination with which teachers and educators in the US have sought out collaborations with neuroscientists. His account recalls a quite distinct trajectory in the development of educational neuroscience in the US when compared with activities in the UK. According to Sousa, while there have been important scientific developments in neuroscience, particularly brain imaging technologies, the real spread of neuroeducation was determined by educators themselves. This often started, as in his personal case, with a 'love of science and [...] passion for teaching' (Sousa, 2010, p 11), but rapidly spread by word of mouth, professional development opportunities for teachers, professional journals aimed at teachers, national education conferences and the advocacy of brain-based teaching and learning methods by educators such as Geoffrey and Renate Caine, Eric Jensen, Robert Sylwester and Patricia Wolfe (Sousa, 2010, p 14). Sousa clearly situates the advancement of neuroeducation within the development of psychologically informed teaching models such as learning styles and multiple intelligences, although he points out that the links to brain research in these cases were limited. There was criticism of a tendency to move too rapidly from findings in the neuroscience lab to educational application in the classroom. However, just as in the UK, since 2000, the US has seen a growth of academic interest in linking neuroscience and education, with programmes at several universities such as Cornell, Dartmouth College, Harvard, the University of Southern California, the University of Texas at Arlington and the University of Washington (Sousa, 2010, p 22).

In 2009, The Dana Foundation, a philanthropic organisation that supports brain research and initiatives intended to educate the public about their brains, published a report on the implications of educational neuroscience for teaching and learning in the arts and creativity. Hosted by Mariale Hardiman and Susan Magsamen from the Johns Hopkins University School of Education, the summit held at the American

Visionary Art Museum in Baltimore aimed to help educators design classroom activities in the arts to improve cognition (Hardiman et al, 2012). The report notes some of the difficulties of engaging in interdisciplinary conversation, suggesting that in the US context, it has been educators who have sought out relevant brain science research. As Johns Hopkins neuroscientist and founder of the 'Learning and the Brain®' conference, Kenneth Kosik, remarked:

> Educators are seriously interested in research; they are hungry for information. Neuroscientists are typically less interested in education; they haven't gotten into the trenches with educators. Conference participants want to know what they can do when they get back to their classrooms. (Kosik, cited in Hardiman et al, 2012, p 4)

The 'Learning and the Brian®' conference has been run since 1997 by a company called Public Information Resources, Inc, which is now co-sponsored by Harvard, Yale, Massachusetts Institute of Technology (MIT), Stanford, Berkeley, the University of California, Berkeley, the University of Chicago, Johns Hopkins University, The Dana Foundation and others. Holding several conferences, seminars and summer school training events per year, the organisation also has an online shop selling conference recordings and a 'neuroscience and teaching' themed t-shirt. In addition, they convene an award for 'Transforming education through neuroscience'. Other such conferences have also been a regular feature of the promotion of neuroeducation in the US, as Sousa (2010) noted. The Learning Brain Expo is one of the most well known, and has been running for at least 10 years. The conference was founded by Eric Jensen, an adjunct Professor in Business and Education, and director of the Jensen Learning Corporation and his own publishing company, which he runs with his wife. Jensen has authored several books on 'brain-compatible' and brain-based learning (see, for example, Jensen, 2007). The Learning Brain Expo has played a pivotal role in translating educational neuroscience research into classroom strategies and school leadership approaches, and Jensen's company provides many teacher workshops, online resources, DVDs and CDs on brain-based methods, including a brain-based approach specific for teenagers, for children with ADHD, and for enriching the brains of students in poverty, targeted at schools receiving Title 1 public funding for low-income children and low-attaining schools. The highly commodified nature of the services provided here is evident in the saturation of products, workshops, user testimonies and click-through opportunities to purchase them from the Jensen Learning Corporation website.[2] So, too, the branding of Jensen Learning as 'The Original, 100% Satisfaction Guaranteed! Since 1995', and the superlatively upper-case assertion that 'Jensen Learning Brain-based Teaching Workshops Provide Classroom-proven, Freshly-researched, 100% Brain-based Teaching Strategies for the Learning Brain'.

However, it is too easy to be disdainful about the propensity of educational consultants to 'sell' their products and services. Just because they have set themselves up as self-employed or for-profit businesses, this doesn't make their knowledge less 'true'. Some may find it distasteful or politically suspect to claim 'secret' insight into solving the behavioural, attainment and social and emotional problems of children living in poverty, reserving that knowledge to those able to pay for the brain-based solution workshop. However, this commodification of neuroeducational knowledge does not per se make it invalid. Jensen has had to devote some concerted effort to facing off critics, posting extensive rebuttals to one vociferous critic on the Amazon website at jensenlearning.com. He has faced specific criticism for his early publications in which references to scientific academic literature were sparse, and in the mid-2000s, alongside his own doctoral training, it is reported that he changed tack in his presentational style. As Tokuhama-Espinosa (2008, p 131) recounts:

> Jensen has come a long way from riding the wave of popular brain fad to becoming a serious scholar and contributor to the field. Unfortunately, his reputation in the early years branded him a "brain-based consultant", which in many neuroscience circles equated with "light" readings of the evidence, overgeneralizations, and the promotion of misconceptions.

For Tokuhama-Espinosa (2008, p 131):

> Jensen's personal evolution parallels the entire field's maturation. More and more early communicators in the field realize that the good of the developing discipline depends on the accuracy of their work, not just the ease with which their books can be read by average teachers seeking guidance.

---

[2]   www.jensenlearning.com/

The development of educational neuroscience, or 'mind, brain, education science' as Tokuhama-Espinosa prefers, can only be understood in light of this complex jostling for legitimacy which has been provoked by the co-evolution of the academic science of learning and the proliferation of publications, products, consultants, conferences, initiatives and services that have been promoted in the name of brain-based teaching and learning. This is why there has been so much effort put in by educational neuroscientists to debunk neuromyths, to claim authority in communicating neuroscientific research findings of relevance to learning, and to establish the grounds for evidence-based teaching practice. Yet however respectable and professionalised the field has become, with its government-supported research initiatives, academic teaching programmes, journals and learned societies, a certain cultish zeal still surrounds some activity of neuroeducation, and there still appear to be lucrative profits to be made out of the brain culture suffusing the school sector in countries such as the UK and US. I reiterate that it would be disingenuous to over-claim the extent of this educational brain culture, given that relatively little is known about how widespread teachers' enthusiasm (or indeed scepticism) for neuroeducation has become, or how profitable the brain-based publishing industry is. Instead, what can be offered is an exploration of the high political stakes at issue in the re-imagining of teaching as a learning science, and the reduction of education to improving cognition and increasing attainment.

Teaching the 21st-century brain

Neuroeducation has emerged in the specific geo-historical context within which the political, cultural and social rationale for education is being constantly reworked. The dawn of '21st century education' in the UK and US alike has seen education re-imagined as the development of core skills in 'learning to learn', life skills such as flexibility, adaptation, innovation, creativity, communication, problem-solving and self-management. These are the skills said to be required to prepare young people for 'working in a diverse cultural and social environment' and within media- and technology-saturated contexts (DfES, 2006; Hardiman, 2012, p 2). As neuroeducationalist Mariale Hardiman puts it: '[a]s we redefine American education to embrace the concept of 21st-century schools, the emerging field of neuroeducation can play an important role by focusing educators on *how students learn* rather than on merely *what they learn*' (Hardiman, 2012, p 2, original emphasis). Furthermore, both within the UK and US, educational achievement, league tables and performance measures (relating to individuals, schools, school districts/local authorities and nation states) have become perhaps *the* central driver of educational policy, with education reimagined as a source of global economic comparative advantage. It is these conditions that make neuroeducation an attractive prospect for educators operating within a discursive framework in which there is a repeated promise to improve, accelerate, enhance and increase the quantity of learning.

The development of brain-based teaching in the US has had a significant impact on teachers' encounters with neuroeducation. Here I give just two brief examples: the Learning Brain Europe conference, and the Brainology programme. The Learning Brain Europe conference evolved specifically out of a group of teachers' experiences of Jensen's Learning Brain Expo in the early 2000s. The teachers from England were so impressed by Jensen's conference that they set up their own version of the conference, and invited several US-based educationalists to speak and provide training for UK teachers on active learning strategies and management tools. One teacher involved in the Learning Brain Europe conference demonstrates a keen enthusiasm for the neuroeducational approach promoted by Jensen:

> '... he brought together neuroscientists who were working in the labs with the rats and, you know, doing serious work, as it were, presenting their findings and then educational strategists, taking those findings, building classroom strategies and models and so and selling them in the marketplace, so it was a fascinating combination of neuroscience and strategies and, you know, management techniques and learning tools and so on.' (secondary school headteacher, interview, November 2013)

One of the main learning strategies showcased by Learning Brain Europe since around 2009 has been the Kagan strategies, which are dubbed 'brain-friendly' teaching methods. Kagan is an educational publishing and training company run by husband and wife team Dr Spencer Kagan and Laurie Kagan. The teacher describes the background to these strategies:

> '... he's a Stanford PhD educational psychologist by training, but he developed and copyrighted this model of cooperative group work, which was based on sound neurology, you know, which parts of the brain are working when children are listening in a passive mode, which parts are working

when they're talking, which parts are working when they're receiving and giving information verbally and so on. And the PET scans, you know, show exactly what's going on and, you know, when you see that, it's obvious isn't it, that children are working hard cognitively if they're working in structured groups.[…] So our, you know, considered response to all that was "well this is the model that we'll adopt". So, over time, we then systematically invested in that as a classroom management model and learning tool and we systematically trained teachers in that and we sent people over to the States to five-day immersion programmes at the Kagan Academy that he runs, we trained our own trainers so that we could sustain it when good people moved off.' (secondary school headteacher, interview, November 2013)

An important part of the appeal of the Kagan strategies was therefore its neuroscientific background, the PET scans that 'show exactly what's going on'. Although billed as brain-friendly, the Kagan methods are essentially strategies for organising communication and collaboration in the classroom, for getting discussion going, and encouraging active group work rather than passive didactic approaches. Nonetheless, the scientificness of the methods is an important selling point, as indicated by their website which states that 'Kagan Structures are scientifically research based as well as backed by classroom evidence from districts, schools, and teachers experiencing success with Kagan.'[3] For this teacher, the research that had gone into demonstrating the impact of such programmes on learning was crucial to their whole school adoption of the methods. After a six-week trial of the Kagan strategies delivered by Kagan trained experts:

'… we were able then to do some research that showed evidence of impact, we did student interviews, we looked at progress measures and you know there were graphs and all the things that you would look for. And then those teachers then, of course, then they become advocates because they're sitting in the staffroom saying, you know, "this stuff actually works", you know, it's much more powerful than me standing up in a meeting and demanding that they do it.' (secondary school headteacher, interview, November 2013)

---

[3] www.kaganonline.com/free_articles/research_and_rationale/

The adoption of various neuroeducational programmes and techniques was testament to the innovative nature of this teacher, this kind of entrepreneurial, risk-taking and yet evidence-informed approach perhaps a key requirement of the contemporary competitive education system. This teacher's school had also experimented with and adopted several other neuroeducational initiatives:

> 'And we've done that with cognitive group work, we've done it with assessment for learning, we've done it with behaviour schools for learning, we've done it with mobile learning technology, you know, we've had projects and research around use of iPads and so on […] we do talk about growth mindsets and we use that as a language and we talk to students about the difference between a growth mindset and a fixed mindset. We dabbled in the Brainology....'
> (secondary school headteacher, interview, November 2013)

Brainology is the second example outlined here. This stems from the work of a US-based (Stanford) psychologist Dr Carol Dweck, whose company Mindset Works® offers Brainology® online products and programmes for both teachers and students. It is backed up by the US Department of Education[4] and has been used in the UK in programmes such as Teach First. Its student version teaches young people about how their brains work, how they learn and how they can increase the strength of the brain, through a 'growth mindset'. The focus of Mindset Works and Brainology is to encourage students to 'focus on improvement instead of worrying about how smart they are.'[5] A research paper by Dweck and colleagues concluded that their progamme showed strong evidence that the growth mindset improved motivation and educational performance. The researchers surmised that while social and environmental factors may be important, these are primarily manifest through the psychology of the child: '[c]hildren's beliefs become the mental "baggage" that they bring to the achievement situation' (Blackwell et al, 2007, p 259).

There are several interesting aspects to this example. First is the value that the company puts on explaining the scientific evidence base for their programme; second is the (by now commonplace in the neuroeducational schema) re-imagining of learning as something to

---

[4]   www.mindsetworks.com/offerings/

[5]   www.mindsetworks.com/webnav/whatismindset.aspx#motivation-and-achievement

be improved and increased because of the impact that our knowledge about the brain can have on our *sense of self*. In an interview with a Mindset Works spokesperson (November 2013), they expressed a clear sense in which neuroscientific insight has given education a new purpose and teachers a new confidence. They stated that new insight on neuroplasticity demonstrates our continued ability to learn throughout life, and that human beings could be divided into two types of thinkers: those with a 'growth mindset', and those with a 'fixed mindset'. Those with a fixed mindset adhere to a more deterministic perspective, seeing themselves or others as particular kinds of learners, and as having been born with a certain amount of intelligence. A typical example of this might be, for example, "I am not good at maths, I am not a maths person." These people might not see the point of continuing to try to learn as they feel that they were not born to achieve in a certain discipline or role. Those with a growth mindset are more open to the idea that our brains continue to develop throughout our lives, and that we can achieve success in education through continued practice. The Brainology programme is precisely targeted at changing students' sense of self; by learning about the plasticity of their own brains they can learn how to develop a growth mindset, and as such can learn better, learn more and perform better in test scores.

Navigating the spaces between brain, mind and behaviour in educational contexts

As has now been noted several times in the inception of neuroeducation as an interdisciplinary endeavour, the translational activities between neuroscience research and educational practice are a crucial arena in which the new discipline has come to be defined. Back in 1997 John Bruer had famously identified the gulf between education and neuroscience as 'a bridge too far', given that there was still some considerable way to go, conceptually and methodology, in even linking the biophysical findings of neuroscience to the mental models of cognitive psychology (let alone implications for practice), and in light of the spurious claims made in the 0-3 early years education movement in the US. But by 2013, if not earlier, Bruer had clarified that his critique was specifically related to the misappropriation of developmental neurophysiology. He reiterated his contention that cognitive neuroscience, which had emerged in the 1980s, had much to offer in shaping educational programmes. Cognitive neuroscience, he stated, can bridge the methodological divide between neural and behavioural data, and retains some modesty around the limitations of functional models and experimental tasks (Bruer, 2013, p 350). In this sense, one of the most prominent critics of neuroeducation, as originally conceived, now seems satisfied that policy dogma has been replaced in many cases by a respectable scientific research programme with the potential to inform and improve educational practice (although he continues to speak out against early intervention policies currently being pursued by the UK government; see Smith, 2014).

Educational Psychologists (EPs) make up a constituency of educationalists who, in many ways, uniquely bridge this fundamental divide between the 'basic' biology of neuroscience and the mental processes to be addressed by the new learning science. Often with professional postgraduate training specifically in cognitive psychology, in the UK, EPs are usually employed by local education authorities to support children and young people who are experiencing learning, social, emotional and/or behavioural difficulties or problems in their educational settings. They may also provide training for teachers on these issues. This section focuses on the perspectives of seven EPs (and trainee EPs who were contacted through their trade union and professional body, the Association of Educational Psychologists [AEP], although their views do not represent this body). The interviewees were based primarily in the West Midlands and Gloucestershire, and the interviews were conducted between October and December 2013. During in-depth qualitative interviews lasting on average an hour, they were asked questions relating to three main themes: (1) their conceptual approach to the relationship between mind, brain and behaviour, and the influence of neuroscience on their understandings and professional practice; (2) their awareness and experiences of the impact of neuroeducation in their encounters with schools; and (3) the importance of contextual factors on their work (including the classroom space, timings of the school day, and socioeconomic background of the schools and children) in shaping the educational experiences of children they identified as experiencing difficulties at school.

What emerges from the interviews is a clear sense in which the professional identities of EPs are changing. They are required to straddle the complex gulf between psychological theory and particular children's experiences of educational settings. With the changing training requirements of educational psychology, more are developing a distinct knowledge base in neuropsychology, and see themselves in various ways as interpreters of and trainers in this field. At the same time they have become arbiters of some of the more dubious claims and educational practices associated with brain culture, and have to navigate a new brain-based language as well as the novel expectations of teachers and parents about their knowledge base. Yet because they deal with children specifically experiencing learning, social, emotional and behavioural problems in educational settings, EPs have an insight into how the theoretical and scientific framing or approach to such problems has an impact on the kind of explanations and solutions available in their work. In this sense, the majority of those interviewed here maintained a healthy scepticism of some of the more reductionist and determinist strands of brain culture, and offered a highly context-sensitive account of the interrelationships between mind, brain and behaviour in specific educational environments.

Educational psychologists' approach to mind, brain and behaviour

In 2006, professional training for educational psychologists was transformed. Before that time they were required to have teaching experience and would take a Master's course in psychology. But since 2006 there has been a requirement to have a doctoral degree from an accredited programme. This shift is a clear indication of the changing expectations of the knowledge base of the EP, away from professional teaching practice and more towards a research qualification which is part university and part placement-based. Some of the EPs interviewed had selected modules in neuropsychology, and one was training to be a neuropsychologist. They demonstrated varying degrees of enthusiasm for the changing knowledge base associated with their role, some clearly embracing the neural turn, and others more sceptical of its explanatory credentials. One recalled an event in around 2009 in which a large group of West Midlands-based EPs attended a day of training in neuropsychology. A trainee described that there was some pressure from teachers for EPs to become experts in neuroscience:

> 'So if there's pressure – I don't know if there's pressure to use [neuroscience], I think there's pressure to know about it. If a teacher starts to talk to you about something in the brain I think they would – I think you would lose credibility if you didn't at least know sort of what sort area in the brain they were talking about or what problem they were talking about.' (EP 1)

One EP had been asked by teachers in her local authority to provide training on the neuroscience of how children learn, and had provided specific training to teachers of children with behavioural problems on adolescent brain development, 'windows of opportunity' for learning, attachment theory and the emotional dimensions of learning. She indicated that she was promoting EPs as a link point between neuroscience and education (EP 2). Another judged that pressure to be fluent in neuroscience as somewhat misplaced:

> 'I have seen educational psychologists argue – putting in writing that, you know, the future of educational psychology is in neuroscience and that it's only a matter of time before every problem that we have to deal with as educational psychologists will be explained by neuroscience and, you know, (a) if that's even plausible, it's – we're absolutely nowhere near it and actually I feel that it's not actually plausible anyway because I think it ignores all sorts of other levels of explanation that I think probably neuroscience can't access.' (EP 3)

His analysis of the future relationship between educational psychology and neuroscience points towards a key distinction that several of the EPs drew between these two academic fields. Just as educational neuroscientists have looked to cognitive psychology as a bridging concept, the EPs felt that neuroscience was not directly applicable in their work or in the classroom, and that it was inadequate as a level of explanation in and of itself, restricted as it was to what this respondent termed the 'algorithm level', concerned only with the computational processes of the brain:

> 'Because neuroscience would be completely ham-strung and be sort of like, you know – if neuroscience was really just the study of the brain, it would be completely ham-strung because it depends on the cognitive psychology and the study of behaviour and things like that, and theories about thinking and how the mind works, rather than the brain, in order to come up with its theories, because it's really kind of – you know, neuroscience is kind of an explanation of – it's sort of like a brain-level explanation of the mind level, but the two things, you know, have to go together.' (EP 3)

The complex interrelationship between the mind, brain and behaviour in specific environments was therefore seen as a crucial corrective to a narrow focus on the algorithmic level of neuroscientific explanation. EPs expressed the view that the messiness of the problems and real-life situations in which they work, along with the feelings and perceptions of teachers, parents and children living within complex environmental systems, has a significant impact on the kinds of theoretical frameworks of mind–brain–behaviour that they operationalise. A few expressed doubt that we would witness any great 'neural' transformation of education, since the neuroscience would always rely on cognitive psychology as a link between brain and behaviour:

> 'But I can't imagine a situation where, say, cognitive psychology ceases to exist as a level of explanation at all, and that there's a direct link between sort of brain and behaviour without any idea of the mind being there.' (EP 3)

Yet despite this appreciation for an integrated mind–brain–behaviour model, EPs were clearly being pulled in different directions by the advent of educational neuroscience. In some cases, their scepticism came from their commitment to prioritising the environmental drivers of behaviour and the importance of a child's experiences on shaping their learning and education. Some were keen to point out that their approach was not limited to the 'within person' model implied by neuroscience, and that their practice was derived much more from the classic educational developmental psychologists such as Piaget and Vygotsky, and psychotherapeutic approaches which have the 'mind' situated in its perceptual environment as a central focus, as the following quote indicates:

> 'I would say I'm systemic and ecological and very behavioural. And I always look at it – we always have this phrase, "within child" and then sort of "systemic environmental" and I'd say I'm not particularly within child. By within child I mean kind of a medical approach, perhaps a neurological approach. But then systemic, I mean more looking at the environment and seeing how it shapes behaviour.' (EP 4)

Conversely, one EP considered the separation of environment and brain in this way as a mischaracterisation of neuropsychology. She felt that the environmental context was a key factor that could be accounted for within the educational neuroscience paradigm. She cited the advent of evidence of brain plasticity and synaptic pruning as core mechanisms for environmental factors to shape the brain and behaviour. Despite this assertion, however, she noted that:

'It's interesting, because we don't use the word "mind" so much at work. I can't remember the last time I wrote or said the word "mind" in a work context.[...] I hadn't thought about that. I talk a lot about brain and behaviour, not mind so much. So there's possibly something in that, I don't know.' (EP 2)

Fundamentally, however, EPs worked within school contexts on problems that they considered to be social, emotional, behavioural, relational and environmental. Many of them therefore believed that a neuroscientific approach would hold little sway within the ecological system of specific schools, specific families and specific socioeconomic environments. The following observation neatly summarises one EP's concerns that the intervention they can offer must relate to people's properly situated stories and not brain scans:

'... ultimately we come to the same place which is "we are never going to know how this child's brain works" because we just don't have the money in the NHS to go measuring every child's brain activity and *we certainly don't know what the brain is doing in situ.* So therefore it may well be physiological but *we've only got social and behavioural context with which to generate interventions.* So I try to go with all the people's stories as much as possible if I need to because it is in the interest of the child [...] ultimately we have to go off the idea that there's a physiological basis for a behaviour because there's nothing physiological we can do to help. We have to still do social and educational interventions.' (EP 5, emphasis added)

To go down this neuroeducational road, it was thought, would take the educational psychology profession back at least 20 years, a time during which "EPs were these mysterious experts that came in, took children off to a room and did something, like a test with them..." (EP 1). Yet if EPs have become more modest in their performance of expertise, the advent of neuroeducation was clearly felt to pose a significant challenge to their status. In order to address this threat to what EPs saw as their insecure profession, those interviewed had considered their changing role as intermediaries between neuroscience and education, either as advocates or as detractors.

Educational psychologists as arbiters of neuroeducation

Although the extent to which neuroeducation and brain-based teaching programmes have infiltrated schools is not known, there is a perception that teachers are readily adopting such initiatives, often in uncritical ways. Certainly, the EPs interviewed recounted many instances of neuroeducation in their local schools, and in some cases, the wholesale adoption of brain-based teaching packages by their local education authorities. Given their evolving role within the education sector, and challenges to their expert status noted in the previous section, EPs have in some instances begun carving out a new purpose in addition to their role of supporting children and young people with difficulties within the education system. Not only did many of the EPs interviewed claim an important new translational role as interlocutors of the new research insights from educational neuroscience, but they also regarded themselves as informed sceptics who should support teachers in understanding the claims made by protagonists of brain-based teaching. In some cases, EPs expressed a stronger sense of their remit, conferring a resistance to the potential reductionism of brain culture. At the same time, they often retained a sense of optimism about the potential value of neuroeducation if it is approached with suitable caveats concerning its blind-spots and its limitations.

When asked about whether they had seen examples of anything that could be described as neuroeducation in schools, all the EPs recalled some experience. The most frequently mentioned teaching examples were Brain Gym® and memory training. Several also highlighted the influence of neuroscience on approaches to children and young people's mental health, particularly relating to novel understandings of the impact of trauma and neglect on brain development, attachment theory and social/emotional development, and ADHD within the education system. The neuroscientific approach to these phenomena was seen to be highly significant in determining the formal practices of EPs in their everyday work. It also shaped the perceptions of parents, teachers and medical professionals whom they encountered. The blurring of the boundaries between mental health and learning difficulties was seen as an important, although by no means unproblematic, development.

Challenging neuromyths

The much-reported use of Brain Gym posed an interesting case for the EPs interviewed. Brain Gym, or educational kinesiology, refers to activities in which teachers use movement to improve cognitive processes. Developed in the 1980s by an educationalist Dr Paul Dennison, and his wife and colleague Gail Dennison in California, Brain Gym proposed that students considered 'learning disabled' could be helped by a set of 26 physical activities reminiscent of the kinds of integrated and spatially aware movements usually experienced in the first years of life.[6] On learning that Brain Gym was being used in hundreds of UK schools, a group of scientific societies wrote to local authorities in 2008 to warn them of the lack of scientific evidence for Brain Gym, and there was much media commentary on this affair (Randerson, 2008). Some EPs judged it to have been 'rife' in particular local authorities where they had previously worked, and most had heard of it or versions of it being used in their schools. However, the EPs' attitudes to Brain Gym were not as simplistic and dismissive as might be expected, given the public controversy surrounding its pseudo-scientific credentials. Some regarded the research on Brain Gym to be mixed, highlighting that any kind of exercise break would be good for learning. While there may be no evidence for its effects on cognitive processes, several EPs believed that it would do little harm. But at the same time, they wanted to warn schools against buying into such programmes, paying for expensive training or licenses, and were wary of the use of neuroscience and the brain in marketing educational products:

'I don't know too much about [Brain Gym] as an intervention but I know that it's not particularly evidence-based. I think it's a really good example of someone making a great company out of a little bit of research! But from my point of view, again as an applied psychologist, we always want to put the best interventions in and we want the best evidence base for them and the thing with Brain Gym, it's kind of one of those things, you know what, it's probably not going to do any damage to the children at worst, it's just exercising. But everyone on my course, if you go into a school and the teachers are like "oh, we're doing Brain Gym" and there's a bit of a sigh inside, like oh, okay.' (EP 4)

Another aspect of neuroeducation identified was memory, concentration and cognitive training, for instance, packages such as CogMed, a computer-based programme for improving attention, which is marketed to a wide range of target audiences including

---

[6]   www.braingym.org.uk/about-edu-k/the-founders-of-edu-k/

'children and adults with attention deficits or learning disorders, victims of brain injury or stroke, and adults experiencing information overload or the natural effects of aging.'[7] Two of the EPs were sceptical of the claims made by such products and services, stating that although there may be some research that shows that concentration was improved for these specific tasks, there was no evidence that there would be any longer-term improvements in attention and memory. The EPs thus saw their role as countering some of the over-claims which are made in the marketing of such products to teachers and parents, and to bring their fidelity to the real evidence and neuroscience to bear in their advice:

> '... these things then get generalised when research gets out there and there's a lot of money behind some of these [memory training programmes]. For me, I think it's a lot more helpful to work with the individual teacher, the school, on how they can improve their general classroom environment than buying into a specific programme. Also, if we think about how the brain develops and how the neurons fire and fuse together and eventually the connections get stronger, that's all about daily habits rather than doing something for an hour a week. Doing something for an hour a week, that doesn't actually go along with neuroscience. Neuroscience is about doing something repeatedly and repeatedly until it becomes stronger and more solid.' (EP 2)

Translating neuroevidence ... sceptically

As will by now be apparent, EPs appear to have a somewhat ambivalent attitude towards the role of neuroscience in education, and particularly with regard to their work with children and young people experiencing difficulties within the education system. Some saw neuroscience as being rather unhelpful in terms of actually providing guidance for their professional practice. One EP attended a teacher training session on attachment theory during which several references were made to neuroscientific evidence. Attachment theory refers to the work of psychologist John Bowlby, which was developed in the 1950s and 1960s, focusing on the importance of the relationship between a primary caregiver and child, and the damage to social and emotional development when that relationship is broken, problematic or separated.

---

[7] www.cogmed.com/users/

He speculated that the neuroscience was mobilised during the training in order to convince teachers of the right course of action in dealing with children with challenging social and emotional behaviours, even though no direct link was made between the neuroscience and any strategy for action:

'But then when it gets to the point of connecting in the images of the brains on the screen which are really powerful and you know bar graphs and all these big, posh words that most teaching assistants – as that was the majority of the people in the audience – wouldn't know meant – it's a really powerful, discursive technique to get people to think "this is right, this is the way to do it". But what in the attachment sort of training, conference thing, the purpose of the brain being there was for – I don't know. I couldn't really see the purpose. There were good strategies suggested but that was sort of separate from the brain.' (EP 1)

Another expressed a similar concern about the recent neuroscientific labelling of some children as 'callous/unemotional':

'… maybe that there is a kind of a neurological basis to that that can be identified, but what's the implication for practice? I mean, it's useful to have identified it, but, you know, in terms of practice, it's going to be – you're going to have to find some way of working with it in a school on a sort of behavioural/cognitive/emotional level.' (EP 3)

The main issue for EPs therefore seemed to relate to the difficulty of translating neuroscientific evidence into workable solutions in educational practice. The identification of neural markers for learning and behavioural difficulties was not seen as useful, and indeed, could get in the way of their relationships with teachers, parents and children. It was a step backward in terms of the efforts EPs have made in supporting their clients through consultation rather than through neuroscientific expertise, and through developing long-term solutions rather than short-term answers:

'I think it's the short answer thing. I think there's a legitimacy to it. So I've seen health psychologists – no, not health psychologists – clinical psychologists who are trained in a similar way to ourselves but they work for the health service and they will be – their role will be – or their department role will be neuropsychology – but they'll do the exact same test that we would do. When a teacher has got a neuropsychology [assessment] – but it's really only an IQ test called a neuropsychology assessment – then they can take that and go in all kinds of directions with how absolute it is […] but we send that message – it says neuropsychology assessment at the top and it's only the fact that I know the psychometric test that's been used and I know it's a paper and pencil test and no brain measure has been done at all. But they can't possibly know that we're telling them, "this is a neuropsychology assessment", so they think something has been done.' (EP 5)

With these caveats in mind, several EPs still believed that they could have a potentially pivotal role in the dissemination of neuroscientific evidence and its application in practice, a role which could 'revolutionise education', the curriculum and understanding of children's needs, by helping teachers to distinguish neuromyth from 'neuroevidence', and by helping teachers to interpret scientific evidence, to which they had little access. It was noted that time-pressed teachers were much more likely to do a quick internet search on brain-based activities than they were likely to access original scientific literature (often hidden behind journal pay-walls), or have the detailed knowledge required to interpret those studies in the context of broader research evidence. What was required, according to one EP, was to "flood out with the myths with factual information" and to empower teachers to demand a strong evidence base for all their practice. This would require, they asserted, a responsible attitude to promoting neuroeducation, and a sceptical stance towards what could and could not be explained and/ or legitimised by the science of learning.

As such, EPs saw their scepticism as a key attribute of this translational role between brain science, the mind and learning behaviours. For some, this meant being reserved about the state of neuroscientific knowledge of relevance to education, with one EP regarding neuropsychology as a "baby science" (EP 6) and another specifying its reductionism as a problem:

'I don't think we have the tools at the minute to measure everything there is to know in the brain or the mind. I think it's quite reductionist to reduce, sort of – and I think most EPs from a social standpoint, to reduce human experience down to, sort of, single cell firing or quantum substrates and things like that, wouldn't be – it wouldn't sit right in that there must be more. I would approach, sort of, neuroscience in, sort of, educational psychology with trepidation and a good deal of healthy scepticism.' (EP 1)

Others, as has already been mentioned, were sceptical as to whether experimental neuroscience in the lab could be translated into logical and rational forms of practice in real settings, particularly those EPs who did not want to limit themselves to what they deemed to be medicalised 'within child' explanations for learning difficulties. But so, too, the language used in neuroeducation poses a problem for EPs because of the way it can disempower teachers, parents and children. Noting again something of a revival of enthusiasm and neuroscientific evidence for attachment theory, one EP described how one practical solution to a child fearing abandonment had been promoted to him during a training session on attachment. If this child, who may have experienced neglect or separation in the past, wanted the attention of the busy teacher, what they could do was give them something, such as a board rubber, to hold, so that the child would be reassured that the teacher would come back to them. This idea really came from psychoanalytic concepts around transference objects that had hitherto not enjoyed the evidential status of neuroscience. But now, this EP reported somewhat cynically, there was neuroscientific proof! The problem for him was that this proof would have no bearing on what advice he would give to teachers, and that, to use the language of neuroscience, would be disingenuous:

'But how they apply day-to-day I don't think – I don't know how – what their messages are about the brain and the negative effects of early childhood experiences how they impact on my profession or what I would suggest to a teacher. I really don't know – you know I would never write "I'll give them a transference object and dorsal lateral cortex will increase in size by whatever per cent over a period of time" – it just doesn't – you would never – I don't think you'd ever write about that or it would never come into the conversation.' (EP 1)

Resisting neuroeducation

While the majority of the EPs interviewed maintained what they deemed to be a healthy scepticism of neuroeducation, judging its potential value to outweigh its limitations, others were more vocal in their resistance to brain-based teaching and the more academic educational neuroscience alike. They expressed specific concerns about the potential determinism, reductionism and medicalisation of neuroeducation:

> 'I think, you know, I think it would be an EP's job to guard against some sort of, you know, call it brain culture, to guard against that because there's an increasing prevalence of, sort of, neuroscientific language going into inset training for teachers about things like attachment, about things like literacy difficulties, about things like behaviour and emotional and social disorders and some of it, sort of, sitting there as a trainee as well, sort of, in the same courses as some of the teachers – it's quite scary because I think when one starts to talk about the brain there is a chance that unless you have a good grounding in, sort of, what comes before you get to the conclusions about neuroscience, you can start to think that the brain is fixed and immutable and it can't change [...] we run the risk of really over-medicalising everything – sort of, there's a big movement in the EP, sort of, world to fight against like this medicalisation of childhood, and things like happiness and things like social and emotional aspects of learning and emotional intelligence.' (EP 1)

Even if neuroscience has made concerted efforts to theorise, model and evidence the plasticity of the brain and the importance of environmental factors in brain development, there is still a sense in which neuroscientific explanation in policy and practice can marginalise social and ecological accounts of and solutions to the kinds of learning difficulties faced by many children in school contexts:

> 'It's not explicitly there. But then if you're coming from a social perspective like a lot of EPs, then it could be quite hard to change views about a young person or about their trajectory from a social perspective if you're fighting against quite [a] powerful sort of biological discourse.' (EP 1)

In this sense, several EPs regarded neuroeducation as politically significant in changing power relations between teachers, professionals, parents and children, and regarded the advent of the language of brain culture in schools as something to be questioned:

> 'People are going to have a lot of clout in professional situations in children's lives to say this is the way this is, this is the way this child is. I think the impact is huge.' (EP 4)

Accounting for context and governing norms

One way in which EPs have sought to address the political issues at stake within a neuroeducational discourse is to maintain sensitivity in their work to the importance of context in shaping learning experience, to approach the learner *in situ* within their educational system, and to propose solutions to problems that they regard to be ecological. It is perhaps because these EPs regarded learning to be an inherently social phenomenon that they were resistant to the re-imagining of learning in neural terms. The EPs interviewed had clear views on the importance of context to learning, including the immediate context of the classroom environment, the temporal context of the school term and the wider socioeconomic context of the school. Several mentioned their awareness of research that showed how the level of visual stimulation within classrooms could affect learners in different ways, for instance, by over-stimulating those with attentional difficulties. Others were interested in how the layout of a classroom affected teacher–pupil relationships and behaviour, although they regarded evidence on this topic to be scant, and for schools to be faddish in terms of chopping and changing their layouts.

They also mentioned how little training was given to EPs on the impact of the built environment and environmental psychology approaches to learning, considering this to be a key limitation to their knowledge. At the same time, they understood how schooling itself was a very specific context for learning, and that their encounters with schools at different times of the year was very revealing in terms of who and what got characterised as a learning 'problem'. The number of referrals they got at the start of the school year was relatively large, when, as one EP put it, "teachers are back in school, they're on it, they've had the six weeks break, you know, they're ready and they're thinking pro-actively because they're not tired out yet" (EP 4). But as the term got nearer to Christmas, and particularly summer holiday periods, they noticed that their referrals slowed down dramatically, partly because schools may have used up their budgets, but also because teachers felt that they could just hold in there and cope until the children moved on to another class the following year.

But the main determining context for the EPs was the socioeconomic context of the school, including the expectations of the teachers, the school ethos, and the communities and families from which the school students came from. One EP with experience of both rural and urban schools recalled that schools in different contexts have very different standards of acceptable behaviour, which has an effect on what are seen as behavioural norms:

> 'In my old school we had a good laugh in the office when one of the psychologists got a phone call – again, from a really lovely rural school, very middle class – and they were like "We want to refer Johnny to you" and it was like "Oh, what happened?", "He swore in the cloakroom" and we were like "Brilliant, send him off!" [laughing] We just couldn't believe it. We were like no, that's not what we deal with.[…] Yeah, that actually happened. I think that's why we laughed because you've got to laugh. But it also makes a difference that […] ADHD has more prevalence in children from a [deprived] economic background and yeah, I definitely find that in my work.' (EP 4)

So children deemed problematic at one school would not be judged so at another school, or even with different teachers within the same school. The EPs recounted several instances where children had been perceived as having significant behavioural problems with one teacher. These problems seemed to disappear when being taught by other teachers, or in a different classroom. Equally, children managing their behavioural difficulties very well in primary school could face serious problems moving into a new context in which the teachers and school hadn't developed strong relationships with them. Conversely, it was also thought that students in more socioeconomically deprived catchments may be less likely to receive an ADHD diagnosis or an EP referral, something an EP described as the 'South of the River' issue in one Midlands city:

> '… there's this thing called "South of the River" because that's where the social deprivation is, so south of the river primary school, 400-odd children. Obviously this is an assertion, it's not based on fact, but I think if you'd been

in that school you wouldn't have even been referred to our service because they have so many other children who are worse off. It would have been such a different context.' (EP 4)

The correlation between behavioural/social and emotional problems and socioeconomic context was regarded as strong, and EPs saw themselves as having an important role in addressing children's problems in light of their contextual experiences, although they often did limited work in the home or with families. But their educational interventions were shaped by context in different ways. For some EPs this might mean that in areas of poverty they might focus their work on fostering home–school links, and encouraging parents with low levels of education to get more involved with schools. EPs often played a role in helping teachers themselves to understand the root causes of children's problems, which may have been more to do with their readiness to learn than their innate learning abilities:

> '... the children that we would work with, it's all well and good going into a school and a teacher saying "we need you to do an assessment on this young person to find out how we can make them learn better", but taking a step, sort of, before that, I think we're always asking the question – or from my personal practice, asking the question, like, "is the young person even ready to learn?", and that is where the social side of it comes in in the sense that you know what is the young person's home life like, their previous experiences, what are they going through now? Most of – I would say from my experience so far – most of what is involved in getting a young person and their environment to change with regards to their learning is a lot about relationship management and negotiation, and giving the adult working with the young person new insight into that young person.' (EP 1)

Meanwhile, several EPs saw the learning difficulties of particular children and young people to be more inherently social, cultural and systematic than neurological or developmental. In other words, they prioritised the environmental or contextual explanations of difficult behaviour and learning problems, noting how the two have often become conflated. A few mentioned the pressures put on schools via league tables to prioritise attainment over wellbeing. Others speculated whether behavioural diagnoses such as ADHD actually said more about our society than our brains, given the specific cultural, technological and economic contexts in which young people are growing up and are schooled:

> 'You have the discourse that every young person is an individual with individual rights and a right to have an opinion and express that opinion. At the same time you have a whole sort of societal system that now promotes, like, constant engagement in everything. So whether that be connected all the time online, TV, games, all that sort of thing – and yet when you put them into a classroom with this sort of constant activation and this idea that you can express your opinion because you're an individual human – and you put them in the classroom and you say, sort of, "Okay, be quiet, sit down, don't talk, listen for an hour, don't move and do your work" and it's sort of, like, but you're giving these young people all this stuff and then you're putting them in an artificial situation because culture isn't like that anymore, basing it on, like, an education system which is a good couple of hundred years old, and yet you're saying there's a medical problem here when they can't sit still and where they want to talk.' (EP 1)

It is this broader context, this EP argues, which leads to the medicalisation of behavioural and learning problems which are actually context-specific, and solutions which are based on improving parents' and teachers' relationships with each other. He describes how "we get a lot of pressure from parents to push for diagnosis because a diagnosis for a parent is safe and it's reassuring" (EP 1). A diagnosis of a developmental disorder such as ADHD does not take place in a vacuum, as many researchers have already demonstrated. Some of the EPs interviewed appeared to resist this medicalising discourse, and were minded to understand ADHD instead in the context of changing educational and behavioural norms, and the specific geographical contexts of particular schools.

The identification, negotiation and targeting of *learning differences* is a theme which has been the focus of considerable attention of both the research of educational neuroscientists and the EPs who act as arbiters of neuroeducation. The implications of neuroscience for developmental disorders such as ADHD was brought up several times by the EPs interviewed as an example of the specific political issues at stake within a brain culture in schools. ADHD connects the sometimes apparently disparate worlds of complex scientific and philosophical debate over mind, brain and behaviour with public controversy over neuroeducation (in this case, brain-targeted pharmaceutical treatment), and the context of the emergence of neuroeducation in some interesting ways.

The first thing to note is that while there might be neuroscientific research currently underway on ADHD, its diagnosis is based on a questionnaire which is filled out by teachers and parents, not on biological or neurophysiological markers (the Vanderbilt assessment scale; see Wolraich et al, 1998). This has led many to question the widespread use of a psychoactive drug, Ritalin, in its treatment (see, for example, Graham, 2010), since there is no adequate model or empirical evidence of the link between the subjective perceptions of a child's behaviour in the specific environment of the school, and the neurophysiological processes of the brain. In other words, the child's brain and behaviour have been conflated without recourse to the mind or mental processes, that is, how we (culturally) perceive something such as attention, its lack, and hyperactivity. As one EP asserted:

> 'Again, going back to ADHD and looking at the neurological/biological basis for traits such as hyperactivity, well it might sound too ridiculous, but what is hyperactivity? You've got to put a parameter on that concept....' (EP 4)

The majority of the EPs brought up the topic of the over-diagnosis of ADHD, and offered alternative accounts of this disorder, including teachers' perceptions, the socioeconomic background of families, relationships of abuse and experiences of neglect, and the promotional efforts of pharmaceutical companies. They felt that the reductionist labelling of children in this way was not helpful in terms of offering workable behavioural solutions to their educational problems, and that the bottom line was that *teachers' attention* in schools was highly conflicted between individual learners with different needs. While many could see that medical treatment of ADHD was a useful option, and understandable to a degree, they felt that the disorder was over-diagnosed, and furthermore, that it followed a clear correlation with socioeconomic class. Some felt that the way in which children were treated at school as a result of such a diagnosis could be an unwanted side-effect of the ADHD label. For many, it was a good example of the complex interrelation of factors that shape learning experiences for which a neuroeducational approach can add little insight:

'I think the important thing with ADHD, and the important thing really in the role of educational psychologists, and that's probably one of the reasons why I like educational – the profession so much – is that it's got a really strong emphasis on the individual needs of the pupil and that – and the needs of the pupil are determined by so many different factors: family factors – you know, you couldn't list them all, really, but it tends to be, you know, family, school, where they live, economic circumstance – you know, I mean, I could go on and on – and individual factors within the child, to the extent that I find it very hard to see what neuroscience can offer to an individual child in order to help explain and explain their – the difficulties that they're experiencing or how they're different from other people.' (EP 3)

This section has examined the specific perceptions of EPs who have been exposed to neuroeducation in their professional training and development in order to better understand some of their conceptual and practical concerns in adopting neuroeducational approaches in contemporary formal settings. Given their work with students already labelled with learning difficulties, emotional and behavioural problems and psychological disorders, EPs occupy an important bridging role between research and practice, and in processes of *normalisation*. They see their role as the responsible, sceptical and sometimes resistant arbiters of neuroeducation as essential to their evolving professional identities. Their experiences in school environments and their desire to qualify brain-based accounts of learning with social, relational and contextual factors attests to their critical stance towards the role of neuroeducation in the governing of psychological subjectivities and shaping behavioural norms.

Re-politicising neuroeducation

There has been as much, if not more, critique and caution voiced *within* the establishment of neuroeducation as from outside, where the attention of sociologists, philosophers and political scientists on these developing educational trends and insights has been relatively limited. While there has been some debate over the ethics of neuroeducation, relatively little has been said about its politics. Two notable exceptions are the work of Edwards et al (2013) and Wastell and White (2012) on the use of neuroscientific research as an evidential basis for the promotion of early intervention policies in the US and UK. This work reminds us of the gendered and classed politics of family, parenting and social work interventions which posit the neurosciences as the starting point for creating better citizens (Allen and Duncan Smith, 2008) and better neighbourhoods (Shonkoff and Phillips, 2000) in the context of acute structural inequalities. But there has been little attempt to empirically chart the emergence of neuroeducation in a specific geo-historical context, or to investigate the extent or nature of its impact in school settings.

For many critics, the formalised scientific identification of learning difference and the ascribing of behavioural learning disorders is part of a therapy culture through which neural and psychological ways of knowing have colonised schools (Ecclestone, 2012). There is a sense in which there is no longer any adequate source of explanation (for learning, behaviour or human activity) beyond the neuroscientific, which is given the last word as foundational evidence on which to base teaching and learning strategy and practice. One resolute critic of neuroeducation, Jan de Vos (2014, p 4), is concerned that the brain culture in education signifies a: 'neuro-*tsunami*, hailing everybody, both pupil and parent, into the neurodiscourse', and asks, 'what, one might ask, will be the effects if soon all generations become immersed in the hegemonic neuro-discourse?' In the context of brain culture, there is an assumption that teachers, parents, EPs, children themselves, and the politicians and policy-makers shaping their learning environments should think like scientists, that no decision should be taken without recourse to what the neuroscience evidence asserts. It has been asserted that culturally, we live in an era of 'neurological reflexivity', and that the governing of citizens and civil society within this culture must be informed by neuroscience (Grist, 2010). For de Vos, this is a significant misstep because of the way in which neuroscience itself is avowedly de-contextualised. Historians of neuroscience remind us that neuroscience does not recognise the specific (academic, popular, geo-historical) circumstances of its own production. That is, the neurosciences are entirely reliant on the culturally specific existence of the anthropological figure of the 'cerebral subject' which emerged out of the 18th-century enlightenment epoch in Western Europe, and by which the brain came to be seen to be the source of human identity and self (Vidal, 2009, p 25).

It has been argued that neuroscience is therefore immodest in promoting a foundational, universal and hegemonic account of human activity – it does not recognise its own partiality and specificity. Furthermore, for such critics, there is *no bridge* between the brain and behaviour, and neuroscience cannot be rescued by cognitive psychology, as those who seek to connect learning with the brain through theories of mind would contend. The fundamental problem of applying neuroscientific insight to policy and practice, therefore, is that neuroscience can only study pre-conceived psychological theories, not actual behaviour *in situ*. As de Vos states: 'what is laid under the scanner is not the psyche, but rather, psychological theory' (2014, p 11). There is thus a problem of circularity inherent in the neuroscientific project; it does not study the universal biological human organ of the brain, but the human subject of 18th-century Western modernist science – that very subject that is able to look in on itself from a detached scientific viewpoint. This is evident in the distinctly subjective lengths to which neuroscientists must necessarily go in order to (1) simplify and reduce culturally specific behavioural phenomena to design tasks to be assigned to research participants within experimental situations, and (2) to develop interpretive schema in order to make sense of their data.

Yet this leap is little appreciated within brain culture, where knowing and celebrating the brain has the effect of representing human beings as biophysical animals somewhat out of kilter with our current geo-historical contexts. This celebratory discourse is evident in several brain-based teaching books which start with a basic introduction to brain geography, simplified pictures of the structural and functional architecture of the brain, or even things like 'brain "cell"ebration: far-out facts about brain cells' (Jensen, 2007, p 31). There are also several online resources now committed to educating children about their brains, such as The Dana Foundation's huge resource bank of neuroscience books for children, games and activities on the brain, and its annual 'Brain Awareness Week', to 'celebrate a week of celebrating the brain.'[8] Acquiring knowledge about our brains has become an essential component of becoming an educated citizen. Geographers have noted how children's abilities to manage their brains, emotionally and neurobiologically, has become an increasingly important aspect of the production of educated citizen-subjects through curriculum initiatives such as SEAL in the UK. As Elizabeth Gagen (2013, p 6) notes: 'the desire to educate emotions via neurology has become a vital part of education.' Yet as she effectively describes, the evocation of the neural in emotional education is not simply informed by the straightforward development of scientific rationality and our ever-

---

[8] www.dana.org/kids/

advancing knowledge of the brain. Rather, it is part of a re-working of education in light of the perceived demands of modern geographical contexts: the neoliberal workplace, the increasing complexity of societal and family relationships, and technological change. These have transformed educable citizens from vessels of information to the self-managing and emotionally competent subjects of late capitalism: '[m]odern subjects are expected to know less about things and more about themselves' (Gagen, 2013, p 7).

By elaborating on the cultural and geo-historical context of the emergence of brain culture in education, and by investigating how we come to see ourselves as neural subjects, such critiques point towards the governance and citizenship implications of actually existing neuroeducation, and the re-framing of neuroscientists as the go-to experts of human learning. Education is, of course, an arena in which the shaping of citizenly conduct has always been a core priority. However, the emerging preoccupation with the brain as a material site of governance – not least in the case of ADHD, where the neurochemical management of behavioural difference is conflated with the delimitation of learning norms in the culturally specific environment of the school system – is troubling. Social scientists who in contrast prioritise *context* in understanding broader and longer-term effects of teaching people *qua* brains can offer alternative explanations of learning here. As this chapter has demonstrated, such a role can be played by EPs, who, despite the concerted efforts of distinguished neuroscientists and purveyors of brain-based teaching alike, remain reassuringly sceptical. They foreground context by rejecting 'within child' totalising neuroscientific rationales for learning behaviours within the specific education system, within specific schools, communities and families. They carefully negotiated neuroeducational claims to evidence, and were aware of their limitations and blind-spots.

However, there is still a sense among many of the EPs interviewed here, and within the diverse field of neuroeducation more generally, that the problem with learning is that we don't know enough about *how learning works*. There is an overriding view that the science of learning is simply not adequately developed, but that neuroscience will push away at the frontiers of both the learning process and self-knowledge. This is why neuromyths have persisted, and why educational neuroscience has had to distinguish itself so vociferously from bogus and pseudo-science. The answer, therefore, posed to brain culture in schools is thus *more brain culture*, the consolidation, extension and dissemination of a neuroscientific evidence to improve learning. In this account, it is not only teachers who should be educated in the ways of the brain, but children, too, should know and celebrate their own brainhood. In contrast, what I have argued in this chapter is that learning can never be understood simply as a brain process. Rather, there are high political stakes in neuroscientific explanations of the leaning process in terms of delimiting learning norms and dealing with learning differences in real places. Education is more than the aggregate sum of people learning. The shaping of conduct, behaviour and educational outcomes is a social and cultural endeavour essential to the governing of citizens in specific contexts. The brain of the learning person is not just an algorithm to be optimised.

FIVE

Managing workplace emotions

> Positive psychology is psychology – psychology is science
> – and science requires checking theories against evidence.
> Accordingly, positive psychology is not to be confused with
> untested self-help, footless affirmation, or secular religion no
> matter how good these make us feel. (Nathan, 2009, p xxiii)

Introduction

Happiness, positive emotion, life satisfaction and wellbeing have
become important focal points for UK public policy since the
turn of the 21st century. At the heart of this policy enthusiasm are
developments in the scientific study of these phenomena. In 2005,
influential economist Lord Layard published a book, *Happiness: Lessons
from a new science* (Layard, 2005), which has had a significant impact on
public policy in the UK and beyond. It brought to public attention the
so-called Easterlin paradox. In the 1970s, economist Richard Easterlin
had highlighted that despite ongoing economic growth in the US, the
overall reported happiness of its population had not increased. Indeed,
as others would have it, advanced industrialised nations with the
highest GDPs in the world are suffering from a psychological malaise
of 'affluenza' (James, 2007). In other words, wealth should no longer
be regarded as a guarantor of wellbeing.

Soon after, the UK's Office for National Statistics (ONS) embarked
on an ambitious programme for the development of measures of societal
wellbeing (Allin, 2007), while the OECD held a world forum on how
best to measure societal progress as distinct from GDP. In 2009, an
influential report was published by the former French President Nikolas
Sarkozy's *Commission on the Measurement of Economic Performance and
Social Progress* (Stiglitz et al, 2009), setting out the inadequacies of GDP
measures. The report also recognised that public policy decisions are
most often made in relation to that which is measurable, and thus, that
having solid metrics for the sustainability of growth and wider social
progress would be essential for governments seeking to evade future
(financial, economic, environmental and social) crises. Numerous
think tanks and civil society organisations, such as the New Economics

Foundation (nef) and Action for Happiness in the UK, the Spinoza Fabrique in France, The State of the USA, Inc in the US, and PerCapita in Australia have focused some of their activities on the difficulties of conceptualising, measuring and intervention for wellbeing. In a very short space of time, the new public economics of happiness had gone global, with the first *World happiness report* launched at the United Nations (UN) in April 2012 (Helliwell et al, 2012).

The new measures of subjective wellbeing, often dubbed as part of the 'happiness agenda' (Wheeler, 2014) in public policy, have generated significant debate concerning, for instance, the firmly disputed validity of the self-report questionnaires on which such measures rely (Cromby, 2011), the cumulative effects of a self-regarding therapeutic culture (Ecclestone, 2012), the radical de-contextualisation of the self from the social world, and the realignment of the self with the economic ideals of neoliberal entrepreneurialism and self-management (Binkley, 2011b; Greco and Stenner, 2013). Although happiness has long been a goal of Western governments, the happiness agenda and the psychological knowledges invoked by the ONS have transformed a specific subjective positive emotion into a measurable and comparable object of national governance. In so doing, it has inadvertently shed new light on the damaging psychological effects of neoliberal capitalism, and the forms of emotional, affective, mental and 'immaterial' labour on which it relies. In fact, this contradiction, as Will Davies (2011) neatly summarises, has the potential to open up new avenues of political and philosophical critique:

> One contradiction of neo-liberalism is that it demands levels of enthusiasm, energy and hope whose conditions it destroys through insecurity, powerlessness and the valorization of unattainable ego ideals via advertising. What is most intriguing about the turn towards happiness amongst political elites and orthodox economists is that it is bringing this truth to the fore, and granting it official statistical endorsement. (Davies, 2011, p 69)

The happiness agenda in public policy has clearly been driven not by scientists, but by economists. Meanwhile, psychologists have independently developed the rapidly evolving new field of positive psychology. Positive psychology is most closely associated with ex-president of the American Psychological Society, Professor Martin Seligman at the University of Pennsylvania in the US. Although Seligman and numerous other psychologists have studied the positive aspects of human nature within psychology for several decades, the inception of positive psychology as a more unified and influential movement dates back to as recently as the late 1990s. As a sub-field of psychology, positive psychology seeks to counteract a perceived obsession within its parent discipline with all things pathological, disorderly and negative. It thus turns its investigations to positive emotions and personal character strengths. In the terms of its proponents (Lopez and Gallagher, 2009, p 4), it therefore successfully 'rounds out' psychology. This rounding out, however, has been matched by concerted efforts to distance positive psychology from so-called 'happyology' and positive thinking, as this chapter goes on to explore. Its focus is thus on the eudemonic aspects of happiness, targeted towards an ethic of the good life (functioning), as opposed to the hedonic notion of happiness as the experience of pleasure (feeling). Seligman (2011) has adopted the term 'flourishing' to encapsulate this difference.

The movement of positive psychology has arguably been significantly buoyed by the advent of emotion science (Fox, 2008) and psychological studies of positive emotion (Fredrickson, 2001; Isen 2009). It has searched for new ways to integrate neurophysiological approaches to positive emotion with cognitive psychological ones (Huppert and Bayliss, 2004). From the perspective of the former, there has been an explosion of research in affective neuroscience which has sought to address the way in which emotions might trouble the common distinctions made between mind and brain (Damasio, 1994; Panksepp, 1998), and thus unlock the path towards clinical interventions and emotional self-mastery. Specific neuroscientific accounts of positive emotions have emphasised a dizzying array of neurobiological features implicated in experiences of pleasure and satisfaction, from the limbic system through the ventral striatum, amygdala and orbital frontal cortex (see, for example, Burgdorf and Panksepp, 2006). Within psychology, new psychometric tests have been developed in order to measure international rates of 'flourishing' (rather than happiness or life satisfaction). One such measure is based on 10 features of positive wellbeing derived from defining the opposite of anxiety and depression, as categorised in the *international diagnostic frameworks, DSM-IV* (the fourth *Diagnostic and statistical manual of mental disorders* of the American Psychiatric Association, published in 1994) and ICD (the International Classification of Diseases of the World Health Organization, published in 1993) (Huppert and So, 2011). This international codification further indicates how specific aspects of psychological theory and method are being adopted in national efforts to *govern through the brain*.

While the theoretical and methodological developments of emotion science, happiness science, affective neuroscience and positive psychology are fascinating in and of themselves, this chapter focuses its attention on the specific political and economic implications of governing emotions in the workplace. As others have noted, the contemporary workplaces of the service-dominated economies of Western nation states are the sites at which positive psychology and happiness science merge, in the concerted effort to ensure the economic autonomy and independence of the worker-citizen from the welfare state (D. Taylor, 2011). As such, it is little surprise that the governmental, scientific and economic interest in wellbeing metrics has coincided with the intensified retraction of welfare in neoliberal states such as the UK and US (Stenner and Taylor, 2008). The workplace is now commonly regarded as the route to social inclusion, to mental and physical health, personal fulfilment and citizenship. The personal and psychological capacities of individual workers are an essential component of this project, and an increasing amount of energy is spent elaborating on the emotional management of 'human resources' (HR) and 'human capital' in post-industrial economies.

It is in this context that workplace wellbeing programmes have grown in importance, with private consultancies, not-for-profit organisations and state programmes for workplace wellbeing thriving in the UK. The Department for Work and Pensions has put several policy measures in place in order to improve the wellbeing of the working population and to assist those with health conditions to get back to and to stay in work. They have even developed a spreadsheet that employers can use to calculate the cost of an employee's ill health and make a business case for workplace wellbeing services.[1] Mental health and emotional wellbeing are increasingly seen as important aspects of employee engagement and successful organisations. Positive psychology workplace training programmes have prospered, and management theories and HR practice have enthusiastically embraced this new sub-discipline and its dual promise of improved wellbeing and performance outcomes. Several variations of positive psychology have been manifest in workplace training programmes to date. Approaches include: strengths-based psychometric testing and talent assessments; Appreciative Inquiry (AI) (a 'positive change methodology' which aims to identify an organisation's success, release positive potential and imagine positive futures; see Lewis, 2011, p 34); 'flow' (which refers to

---

[1]   www.gov.uk/government/publications/workplace-wellbeing-tool

immersing oneself optimally in an experience; see Csikszentmihalyi, 1990); and mindfulness practice (a form of meditation derived from Buddhism which aims to help people focus their conscious attention on the present, brought to popularity by John Kabat-Zinn, 2004 [1994]). The knowledge, techniques and infrastructure for *governing the self* in the workplace have become more and more sophisticated, and the brain and mind are increasingly invoked as the target of skilful self-conduct.

This chapter explores the emergence of positive psychology in the workplace as an emblematic feature of a brain culture in which policy and practice become heavily indebted to psychological and neuroscientific accounts of human decision-making, our emotional engagements with the world, and our behaviour. The first section reviews developments in approaches to positive emotion and affect in the HR, management and organisational studies literatures. The second section explores the relationship between the scientific credentials and business relevance of positive psychology. It demonstrates how the pursuit of positive emotion (optimism) and continuous self-perfection (optimisation) are posed as solutions to workplace underperformance and economic recession. The final section examines the way in which positive psychology, both methodologically and philosophically, can decontextualise the person from the specific circumstances within which the movement has found its raison d'être.

The chapter thus sets out to reflect on the geographically dislocated sense of human subjectivity offered by the academic and practice-based movement of positive psychology at work. The universalist search for authentic happiness at work is by no means uncontested, and the chapter explores some of the critiques of these attempts at *normalisation* offered within critical psychology. The analysis provided here is based on in-depth interviews with 21 practitioners of positive psychology, strengths-based approaches and mindfulness who are involved in delivering workplace-based training programmes in the UK, with some practitioners delivering such programmes worldwide. These interviews were undertaken between May–August 2013.

Human resource management, from job satisfaction to positive neuroscience

Positive psychology has emerged as a means by which to re-invent the worker as an adaptable, highly functioning individual, skilled in the management of their own and of others' emotions, committed to the pursuit of self-optimisation, and responsible for their own psychological governance. Clearly being a happy worker in a positive workplace is an appealing goal and a good thing. Yet the pursuit of happiness in a Western geo-historical context can be understood as a specific outcome of the alignment of psychology and economic knowledge. There is an economic imperative to ensure workers are adequately satisfied, but not to the extent that they abandon their role as consumers of non-necessity goods and services in search of further satisfaction. It has long been recognised in studies of economic behaviour that *homo economicus* is a satisficing rather than optimising figure (Simon, 1947), and this tension is to be negotiated by recourse to psychological knowledges about the self.

The way in which positive psychology aligns a worker's intimate and personal attributes, strengths and emotional commitments with that of their organisation and, by extension, the economy, is therefore of interest. It is by no means the case that positive psychology has brought into existence this phenomenon of 'emotional labour', as sociologist Arlie Hochschild (2003 [1983]) described it some three decades ago. Rather, positive psychology has enabled the formalisation and professionalisation of interventions in the workplace management of the emotions, both through its scientific credentials and its extensive deployment by workplace trainers, coaches, management and leadership consultants and wellbeing practitioners worldwide. This section explores the influence of positive psychology on HR, business and management studies research since the mid-1990s. In particular, it outlines some of the main approaches to and principles of the often acronym-laden study of positive organisational behaviour (POB), psychological capital (PsyCap), authentic leadership development (ALD) and strengths, as well as the growing interest in the brain, in terms of the neurobiological positivity of workers and the affective neuroscience of the workplace. Helpfully, the papers in one agenda-setting special issue of the *Journal of Organizational Behavior* in 2009 confirm these key themes, namely, group behaviours in organisations (West et al, 2009); the psychological resourcefulness (or capital) associated with 'organisational citizenship behaviours' such as commitment, transformational leadership and job satisfaction (Avey et al, 2009); leadership and the emotional self-regulation of successful leaders (Hannah et al, 2009); character strengths such as zest, hope and optimism (Peterson et al, 2009; Kluemper et al, 2009; Simmons et al, 2009); and finally, physiological employee wellbeing, in this case, the impact of such work-related wellbeing on cardiovascular health (Wright et al, 2009).

It is worth providing a little more detail on this knowledge base since it sets the scene for the empirical analysis of positive psychology as a manifestation of brain culture in the remainder of the chapter. Positive psychology as an academic movement has a close historical relationship with organisational and HR/management studies. But its evolution is also intimately entwined with its deployment as a set of management practices in actual workplaces. It is not, therefore, a straightforward scientific research agenda, as the chapter goes on to explore. Indeed, the co-evolution of its psychological science and its work-based practice is evident in one of the rationales often given for positive psychology – the need for HR and management practices that take the *whole* person as the focus of their interventions. In one sense it is about treating the workers as much more than a human resource. And yet conversely, it is also about treating the person and citizen as *primarily* a worker. As Michigan-based Professor of Business, coaching consultant (and several times ranked number one influential person in HR by *HR Magazine*) Dave Ulrich (2010) writes in the foreword to the *Oxford handbook of positive psychology and work* (Linley et al, 2010, p xvii):

> ... work organisations are increasingly becoming a primary setting where people may (or may not) meet their personal needs because we spend an increasing amount of time at work, because with technology the boundaries of work and non-work have blurred, [...] because organizations have become a primary social setting for many people, and because work shapes so much of our personal identity.

While this diagnosis of the contemporary dominance of work may ring true, the effect of such an account is to universalise the Western pre-occupation of paid work. As feminists have long argued, to conclude that personal identity is almost entirely bound up with such work serves to devalue other forms of meaningful human action which cannot be reduced to economic productivity and essential labour. Nonetheless, the route to happiness for positive psychologists is through what Ulrich describes as the 'abundant' workplace – organisational settings that focus on 'what is right', not 'what is wrong' (Ulrich, 2010), support workers in identifying their strengths, align the commitments and values of workers to organisational goals, base their strategies not on problem-solving but on dreams and appreciation, have a sense of environmental and social responsibility, include high-performing teams, and take into account work–family issues (Ulrich, 2010, p xix).

Happy workplaces: positive organisational behaviour

Positive approaches to organisational management have a long history. The Hawthorne studies carried out by Elton Mayo and researchers at the Western Electric Company in suburban Chicago in the 1920s mark one of the most famous investigations into the effects of improvements to working conditions. The study began by examining the effects of changing the material conditions of the workplace, such as lighting, and showed that productivity increased. Later, they changed other working conditions such as breaks and working hours, and productivity again increased. Surprisingly, productivity increased yet again when the lights were dimmed. For a subsequent generation of positive psychologists, the 'Hawthorne effect' suggested that it was not changes in working conditions, but 'immaterial' factors, such as the participation of workers in decision-making, group dynamics and increased attention, that caused better positive attitudes and performance (Luthans and Youssef, 2009, p 580). For such authors, this long history points towards the value of focusing research enquiry and organisational practice on positive psychological approaches to human resources.

Yet what these authors don't mention is that the Hawthorne effect is also well known as an artefact of the research process itself. It was specifically the presence and interventions of the researchers as observers which had a positive influence on worker motivation. In these terms, the Hawthorne studies suffer from what philosophers of science have observed is a tendency of psychologists to mistake their research apparatus for scientific instruments (Harré, cited in Cromby, 2011, p 847). In other words, the research set up is clearly an intervention in the social context, and no outside and detached observation is possible. Such studies cannot directly measure human behaviour as a thermometer would react to a change in temperature. Instead, the studies themselves contrive to model human behaviour in ways that can never be fully objective and replicable. As Cromby (2011) suggests, this same methodological error plagues the use of the psychometric tests used in national government's wellbeing measures, as well as the kind of character strengths questionnaires commonly used in positive psychology workplace training programmes.

Other early precursors to the contemporary positive psychology movement are identified by Wright and Quick (2009, p 150), who, in their editorial to the aforementioned special issue of the *Journal of Organizational Behavior*, set out the emerging research agenda for this movement. They note that the enthusiastic organisational studies of the 1920s and 1930s at business schools such as the Wharton School at the University of Pennsylvania were stopped in their tracks by the advent of the Great Depression. They highlight in particular a former colleague of Mayo's, Rexford Hersey, whose 1932 longitudinal studies of railroad repair workers and their lives involved him spending whole days, including time outside of work, with them. His data included the biophysical as well as the psychological aspects of wellbeing, and his pioneering appreciation of the emotional significance of work–life balance and contributions to the 'happy/productive worker thesis' has been noted (Wright and Quick, 2009) (although his discovery of the male workers' monthly emotional cycle remains something of a niche concern; see Dickinson, 1933, p 426).

Happy workers: psychological capital

Positive Organisational Behaivour (POB) is one approach to management that is associated with 'actualizing the potential of human resources in organizations', for example, through 'behavioural performance management' including 'contingent positive rewards, positive feedback, and social recognition' (Luthans and Youssef, 2009, p 581). Its focus is on measuring, developing and managing HR strengths and psychological capacities in ways that improve performance. Such strengths and capacities are similar to those promoted in positive psychology, in that they must be scientifically validated and based on sound theory. They are also distinct in that they are state-like rather than trait-like; they are not fixed character virtues but amenable to intervention. POB is also distinguished from positive psychology in that its goal is more explicitly to quantifiably improve workplace performance and profitability rather than human functioning.

POB is based on four main positive capacities, which are also described in Fred Luthans and colleagues' work as psychological capital, or PsyCap. These assets include (self-)efficacy, hope, optimism and resiliency (resilience), sometimes shortened to the acronym HERO. Self-efficacy refers to a confidence to act successfully – drawing on the theories of social psychologist Albert Bandura. Hope is a positive motivational state suggesting orientation towards future goals – after positive psychologist Snyder. Optimism connotes an explanatory style by which we attribute positive events to 'personal, permanent, and pervasive causes and [explain] negative events in terms of external, temporary, and situation-specific ones' (Luthans and Youssef, 2009, p 582, after positive psychologist Martin Seligman). This optimism helps us to think positively towards a successful future, aids us in mitigating against potential risks, keeps us thinking positively, and ensures self-mastery and internalised control (Luthans and Youssef, 2009, p 584).

Finally, resilience is an ability to bounce back from adverse situations such as conflict or disadvantage, and to adapt and prepare for risk – after child development psychologist Ann Masten (2001).

Happily, for management consultants and workplace training providers, these forms of PsyCap correspond almost seamlessly with common-sense understandings of successful workplace behaviours: a 'can-do' attitude, a positive goal orientation, looking on the bright side, and coping with and adapting to change. Luthans and colleagues report very positive outcomes of investments and interventions in PsyCap, with one example of a 270 per cent return on investment, boosting this lucky high-tech manufacturing company's bottom line (Luthans et al, 2006). Luthans and colleagues also run a publishing company, Mind Garden, Inc, which provides a vast number of psychological assessments and instruments, checklists, inventories and evaluations such as the PCQ (PsyCap psychometric questionnaire) which provides accounts of self-perception, and the PCQ 360 Multi-Rater Report which uses the 360 degree management method to ascertain co-workers' perceptions of a person.[2]

Happy bosses: authentic leadership development

Central to PsyCap and POB is a sense that the workplace and organisational culture itself is the source of positivity, and that much positive action and intervention can be taken to improve positivity for productivity. Such intervention requires positive leadership and performance management of HR, which is described by some in the field of positive psychology in the workplace as authentic leadership development (ALD). In a special edition of the business journal *The Leadership Quarterly*, Avolio and Gardner (2005, p 319) describe ALD as having a heritage in the humanistic psychology of the 1960s and 1970s, namely, in the work of Carl Rogers and Abraham Maslow. It came to prominence soon after the publication of Harvard Business Professor, Bill George's best-selling book, *Authentic leadership* (2003, p 18) which highlighted a crisis in corporate leadership in American culture (think Enron). It identified the need to lead purposefully according to one's values, heart, relationships and self-discipline, rather than in blind pursuit of financial market success. While there is some disagreement as to the specific scope of ALD, authenticity is defined by being true to one's self (not fake), being fully functioning, self-actualising and

---

[2] www.mindgarden.com/products/pcqconsult.htm

relatively autonomous; not conforming the expectations of others and social pressures, and avowedly active in (rather than subject to) the social construction of shared realities.

Authentic leaders are motivated by personal conviction. They are thus characterised by a strong self-narrative – a personal story, moral perspective and psychological capital. They act in accordance with this reflexive sense of self, are known for their self-reflexivity, their awareness of others and of context, and their states of confidence (self-efficacy), hope, optimism and resilience (Avolio and Gardner, 2005, p 321). Such leaders use these attributes to shape organisations which value self-development. Their authentic leadership is said to trickle down to their followers, having a positive impact on performance and attitudes (Luthens and Youssef, 2009, p 584). In their analysis of the distinctiveness of ALD from other leadership theories, Avolio and Gardner (2005, pp 330-1) conclude that it has much in common with other leadership theories such as transformational, charismatic, servant and spiritual leadership, but that there are clear distinguishing features. Being visionary and charismatic are not essential components of authentic leadership. Authentic leaders lead through the cultivation of self-awareness by personal example and the shaping of positive organisational environments, not through rhetoric, inspiration and impression management. Authentic leadership shares with servant and spiritual leadership theories a focus on self-awareness and the values of 'integrity, trust, courage, hope, and perseverance (resilience)' (Avolio and Gardner, 2005, p 331), but it is said to be much more highly developed theoretically and empirically.

Overall, POB, PsyCap and ALD are all considered essential components of a positive psychological approach to HR management and workplace cultures. There are two further significant aspects of training happy workers in positive workplaces that can be drawn out from this literature.

Virtuous characters: strengths-based training approaches

The first is an emphasis on the psychometric identification of character strengths, after the influential work of positive psychologists Martin Seligman and Christopher Peterson, who together developed a system of classifying what they assessed to be universal and timeless character strengths and ways of measuring them (Peterson and Seligman, 2004). Strengths are described as 'the subset of personality traits on which we place moral value' (Peterson and Park, 2009, pp 26-7). They are therefore those aspects of personality which can be described as virtuous. In this sense, we can detect a shift in emphasis in positive psychology from experientially positive states to morally virtuous traits. Peterson and Seligman's methods in developing their strengths taxonomy involved extensive and wide-ranging literature reviews, including surveys of psychiatric, youth development, philosophical and psychological literatures, as well as a review of some major religious texts. They examined the character strengths of historical figures said to be paragons of virtue, such as:

> Charlemagne and Benjamin Franklin, contemporary figures like William Bennett and John Templeton [whose philanthropy has funded research and prizes in positive psychology], and imaginary sources like the Klingon Empire [from the Star Trek TV series]. Also consulted were virtue-relevant messages in Hallmark greeting cards, bumper stickers, *Saturday Evening Post* covers by Norman Rockwell, personal ads, popular song lyrics, graffiti, tarot cards, the profiles of Pokémon characters, and the residence halls of Hogwarts [from the fictional book, Harry Potter]. (Peterson and Park, 2009, p 26)

This presumably vast collection of virtuous personality traits was narrowed down by assessing the strengths according to the extent to which they were: ubiquitous/universal; fulfilling; morally valued; not likely to invoke jealousy in others; clearly paired with a negative opposite; trait-like; measurable; distinctive; evident in paragons and prodigies; missing in some people; and the focus of institutional or social efforts to cultivate that strength (Peterson and Park, 2009, p 27). By a process of elimination, a list of 24 character strengths was identified, and a means to measure them (primarily a self-report survey) was developed by the VIA® (Values in Action) Institute, a not-for profit organisation set up by Seligman and colleagues in the early 2000s.

The classification is organised into six virtue groups: wisdom and knowledge; courage; humanity; justice; temperance; and transcendence (Peterson and Seligman, 2004; Peterson and Park, 2009, p 28). The strengths were further measured and tested through focus groups, interviews, content analysis of people's descriptions of self and others, a review or obituaries and case studies of paragon-like figures in order to establish their reliability and validity (Peterson and Park, 2009, p 29). A significant claim of the VIA strengths classification is its global applicability, established through a large internet-based survey which included data from 54 countries and all 50 US states, leading its proponents to 'speculate that our results revealed something about universal human nature and/or the character requirements minimally needed for a viable society' (Peterson and Park, 2009, p 29). One significant problem with this range of methods is, of course, that none of them are measures of what people are like. Rather, they are measures of what people *think* they are like, what people *say* they are like, and what people think they *should* be like. As such, significant social and cultural norms will likely skew the results. Indeed, Peterson and Park (2009, p 30) found that there were some differences, for example, between male and female survey respondents for the 'interpersonal strengths of gratitude, kindness and love' which they simply describe as 'sensible'.

One of the uses to which the VIA strengths classification has been put is to explain the role of these personality traits in determining outcomes in workplace productivity – to refine the happy/productive worker thesis and to examine what kinds of people are satisfied in work. However, the rationale for such research is firmly situated in the aforementioned happiness economics, with studies taking as written that salary increases follow the laws of diminishing returns in terms of wellbeing (Peterson et al, 2009, p 161). Given this notion that better pay doesn't necessarily improve people's wellbeing beyond a certain threshold, Peterson and colleagues (2009, p 161) set out to explore the role of the character strength, 'zest', in work satisfaction. Zest is defined as 'the habitual approach to life with anticipation, energy, and excitement', and connotes a sense of enthusiasm and vigour. They hypothesised that bringing a zestful disposition to work should help to transform that work into a calling rather than a job as a source of money or a career as a source of status. This was found to be precisely the case, based on responses from their 'international' (75 per cent US respondents) online survey of 9,803 individuals.

Strengths-based approaches to workplace training and business management are commonplace, with several different versions independent of the VIA survey available and in use globally by management consulting companies such as Gallup (who use the proprietary StrengthsFinder tool of 34 strengths associated with former CEO of Gallup, Don Clifton), Strengths Partnership (which has a proprietary suite of 'profiling and development' products including Strengthscope), McKinsey and Company (who have integrated signature strengths into their model of 'centred leadership'), and Dale Carnegie Training (who use proprietary talent assessments). In the UK, CAPP is one of many HR consulting companies that use strengths-based tools specifically developed from positive psychology (they use proprietary Realise2 tool of 60 strengths).

The overall goal of all these approaches is to refocus attention on what workers do well, to identify a handful of strengths that they can maximise rather than to dwell on their weaknesses, to develop excellence in specialist competences rather than well-rounded mediocrity. Strength-based tools, talent assessments and psychometric tests are also frequently used in employee recruitment, and have been used in UK government policy experiments by The Behavioural Insights Team with welfare claimants in their jobseeking activities at Jobcentre Plus offices. Their use in the benefits context has met with some opposition in the media (Malik, 2013) and by psychologists (Cromby and Willis, 2014).

Happy brains: organisational neuroscience

The final field of enquiry shaping contemporary HR management and positive psychological approaches to organisational studies is the neurobiology of positive affects and the physiological role of the body in positive workplaces. But a more imprecise trace of brain culture is sometimes evident in the language used by work-based positive psychology enthusiasts and in positive psychology-inflected organisational and HR studies. Sarah Lewis, a chartered psychologist and managing director of organisational consultancy, Appreciating Change, invokes an epigenetic rationale in her explanation of Gallup's focus on talents:

> Talent is seen as behaviour that is an expression of our brain "wiring". As we mature, our brains become uniquely wired as a result of our original genetic endowment and subsequent experience. Any thought or action that we experience repeatedly develops into a well-trodden groove in our mind. (Lewis, 2011, p 44)

The apparently limited plasticity of the brain is further brought in as an explanation for a managerial emphasis on developing strengths rather than trying to fix an employee's weaknesses:

> Their [researchers] argument is that if we haven't developed a particular facility to a certain level by adulthood, we are unlikely to get much return on our efforts to improve. Science backs this up. It is clear that our brains are much

more active at developing and growing new connections before we reach adulthood.[...] The core, we might say, of our unique brain design is in place by the time we are young adults, and the suggestion is that we are better off devoting our energies to working out what we have and how best to use it than trying to fundamentally change it. (Lewis, 2011, p 44)

In addition to these allusions to brain research among practitioners of positive psychology at work, there is increasing dialogue between positive psychology and neuroscience more generally. There have been recent moves towards establishing a 'positive neuropsychology' (Randolph, 2013), and Martin Seligman himself received a US$5.8 million grant from the John Templeton Foundation to initiate a new research programme of 'positive neuroscience' in 2008. This research programme investigates the neural mechanisms of his concept of human flourishing.[3] This is by no means the first foray into the neurobiological study of positive affects. Research by Alice Isen and colleagues since the 1970s has explored how (experimentally induced) happy feelings or positive affects have been shown to enable cognitive flexibility, as manifest in more effective thinking, problem-solving and social interaction (see Isen, 2009, p 504). Indeed, the neuroscience of affects has been growing in importance since the mid-1990s (Damasio, 1994; Davidson and Sutton, 1995; Panksepp, 1998), although the focus is not always solely on the positive side of affects. Others have studied aspects of emotional intelligence specifically in the workplace (Zeidner et al, 2004).

'Neuroeconomists' have investigated the inner workings of the firm from a neuroscientific perspective, for example, identifying the degree of trust in workplaces by measuring employees' levels of oxytocin (Durante and Saad, 2010). More concerted attention has recently focused on 'organisational neuroscience' (Butler and Senior, 2007; Becker et al, 2011), 'cognitive wellbeing' in the workplace and the 'neuroscience of leadership development' (Randolph, 2013). There is certainly a sense of radical change afoot as organisational studies adopt neuroscientific advances in their understandings of behaviour at work, with a celebratory attitude in evidence among some: '[n]euroscience can allow us to finally go inside the brain and investigate these primal causes of behaviour' (Becker et al, 2011, p 934). The establishment of

---

[3] www.authentichappiness.sas.upenn.edu/learn/positiveneuroscience

causation here is clear; the evolved brain, which Becker and colleagues regard to be relatively fixed, pre-programmed and often automatic, is the source of our behaviour, and it is in the neurosciences that the 'most fundamental level of analysis' is to be found (Becker et al, 2011, p 934). For some, the development of organisational neuroscience is more modestly a logical extension of organisational psychology and evidence-based practices in management theory. The advent of brain imaging technologies and of affective neurosciences has established the ground for this new approach, and enthusiasm for this field has already been demonstrated by business leaders (Randolph, 2013, p 103). Randolph notes that much of the work has emphasised the promotion of the brain's executive functions (for example, decision-making, problem-solving) in business executives, in addition to aspects of 'stress and emotional regulation, social interaction and emotional intelligence, and promoting cognition in the workplace' (Randolph, 2013, p 104).

An emphasis on emotional intelligence is also evident in the emerging practice of mindfulness meditation in workplace stress-reduction programmes. Psychologists at the University of Oxford Mindfulness Centre (OMC) have been involved with a spin-off company, The Mindfulness Exchange, which was established in 2012 to provide workplace-based mindfulness courses based on clinical research undertaken at the OMC. The four-, six- or eight-week mindfulness courses share in common with positive psychology the clear aims to improve workplace performance, engagement and resilience to stress and anxiety. The course is supported by a self-help book co-written by OMC founder Mark Williams and *Daily Mail* journalist Danny Penman (Williams and Penman, 2011), *Mindfulness: A practical guide to finding peace in a frantic world*. In his preface to this bestseller, John Kabat-Zinn (2011, p x) highlights the history of mindfulness meditation in Buddhist practice as well as the scientific and medical evidence which has established it as more than 'a passing fad'. The secularisation of mindfulness, and neuroscientific investigations into its effectiveness have been crucial to its current popularity. Several university research centres (Oxford, Bangor and Exeter) have been launched to study the effects of mindfulness and provide postgraduate training in mindfulness based cognitive therapy (MBCT). In organisational studies, meanwhile, mindfulness is increasingly established as a trait, quality of attention or skill of emotional regulation which has benefits for improving employee wellbeing and performance. The integration of physiological accounts of mind–body dynamics and a neuroscientific approach to skilful comportment of this nature has much in common with Thrift's affective geographies outlined earlier in Chapter Two. The bodily, ritual, mystical and therapeutic practices evoked by Thrift as a source of political potential are precisely those put to use within the meditation training programmes offered in such workplace initiatives.

Nothing less than an 'affective scientific revolution' in studies of organisational behaviour has consequently been claimed by some management scholars (Barsade et al, 2003, p 5). They posit radical improvements in the theoretical and methodological sophistication of organisational and HR studies, as well as advances in measurement and in understanding. Basic self-report surveys of job satisfaction have been effaced by this new research programme, and there has been a shift from investigations of reflexive appraisals of working life to studies of immediate and unconscious affective events, the social contagion of affect (for example, collective impression management, attempts to influence others, passing on stress), and the interrelationship between affect and cognition implied by the concept of emotional intelligence. Research on emotional intelligence in organisations, as Barsade et al (2003) recount, is not limited to an individualised understanding of emotion, but is much more attuned to the 'affective cultures' of workplaces, and should be informed by social psychological and social neuroscience approaches to emotion. The work reviewed by Barsade et al (2003) is thus by no means ignorant of social context. What's more, sociological accounts of emotional and affective constructs, which take context (norms and structures) as a determining factor in the expression and experience of emotion, are considered an essential part of this new paradigm. Credit is given to Hochschild's account of emotional labour in *The managed heart* for initiating the affective turn in organisational studies.

Yet it is clear that Barsade and colleagues regard the sociological approach to be plagued by disagreements relating to empirical methods and initial definition of core concepts such as emotional labour. In other words, this sociological strand of research is not scientific enough. Furthermore, the critical intentions of Hochschild's work appear to have been lost in the adoption of her observations and insights into some aspects of management studies. Her attention to the alienating effects of emotional management in the corporate workplace, and her analysis of the root causes of such alienation as fitting squarely with post-industrial capitalism, are certainly not dwelt on in this otherwise emotionally literate organisational research literature. It is by no means a coincidence that Hochschild's work has been only very selectively influential in the field of management and organisational studies, despite its emotional coming of age. Similarly, the mindfulness approach to workplace stress reduction is avowedly silent on the causes of 'our frantic and relentless way of life' (Williams and Penman, 2011, p 2) in favour (no doubt intentionally) of practical self-based solutions. As with the case of neuroeducation, we are incited to know more about our brain than about the world; to train our habits rather than change our spaces. In contrast, a geographical analysis of the association of the science of positive psychology with the business of workplace training and management consultancy in a specific geo-historical context can help us to explain the search for fundamental brain-based explanations for the happy/productive worker in positive organisations, as the remainder of this chapter explores.

The business and science of positive psychology: the emergence of austerity happiness

As has already been noted, positive psychology has become an influential approach to organisational management and workplace training programmes since the late 1990s. It is heralded as a potential solution to global corporate problems such as stress-, depression- and anxiety-related employee absenteeism and 'presenteeism' (being present but ineffective), and dovetails conveniently with an equally global policy emphasis on happiness and workplace wellbeing. Its scientific credentials are a crucial part of this success story, as businesses and organisations seek more evidence-based approaches to employee engagement and HR management. And yet, positive psychology as a sub-discipline has also evolved partly in response to its application in practice, and has been in part driven by a business-led demand for workplace training programmes which simply 'work'. This section examines the co-evolution of the science and business of positive psychology workplace training in order to account for the predominance of its two key goals of producing *optimism* and *optimisation* in the workplace. As such, positive psychology at work is examined as a manifestation of brain culture in the sense that it uses psychological and neuroscientific knowledges in cultivating self-management strategies to promote optimism in the workplace (*governing the self*) and prioritising the pursuit of endless self-optimisation or being 'all you can be' (*normalisation*) as a personal aspirational goal for all workers.

The business

Critic of positive psychology Barbara Ehrenreich (2009) suggests that this academic movement developed in the US economic context out of a business enthusiasm for 'positive thinking' and motivational speaking at a time of acute economic restructuring during the 1980s. Widespread downsizing challenged rational management techniques as businesses began to reorganise around demands of flexibility, technological change, shareholder priorities and global competition. Managers needed innovative ways to get the most out of their workforce, and positive psychology has been mobilised to these ends ever since. Proponents of positive psychology themselves have recognised that the changing context of employment since the 1980s provides a clear rationale for workplace interventions that encourage employees to make the most of an 'unstable situation' in which job security, long-term careers, more intensive, competitive and often disheartening forms of work predominate (Turner et al, 2002, p 717). Practitioners of positive psychology, too, have attributed some of its success to the needs of business managers to 'do more with less' following the global financial crisis of 2008. This economic climate has provided a business opportunity for many workplace training consultants who can provide happiness-based solutions in times of austerity:

> 'I know a lot of organisations got to the point where they had cut as much as they could cut without permanently damaging the organisation, they couldn't ask any more of their people and yet they needed to, so they started to look at what they could put back into their people to rebuild them. That in tandem with the faddish interest in positive psychology and happiness at work and wellbeing and mindfulness has caused people to think that "oh there might be something in this, if we look after our employees better maybe they will perform better". That combination of things has driven positive psychology in business.' (provider of strengths-based workplace training, interview, July 2013)

These contextual factors have been important in providing the business conditions in which positive psychology workplace training programmes have thrived by promising better workplace performance outcomes and cost savings. The business opportunities for positive psychology have been evident throughout its history, with university-based psychologists and management theorists often pursuing commercial sidelines in business consultancy, training services, lecturer tours or proprietary psychometric assessment products such as Gallup's StrengthsFinder or MindGarden's PCQ (Psychological Capital Questionnaire). For some advocates, the business evolution of positive psychology workplace training programmes has many advantages over the staid and long-winded endeavours of the scientific establishment:

> Unbound by academic politesse or the demands of research journal gatekeepers, organizations are interested only in answering a single, simple, powerful, effective question: *What works?* Show me what will make a difference to my bottom line, and I'll do it – irrespective of what theory it has been developed from, or of whose reputation may be challenged in the process. (Linley et al, 2010, p 6)

As one provider of strengths-based workplace training described, the application of positive psychology in business organisations involves the modification and practical adaptation of the academic theories in order to meet business needs:

> 'When we launched [our own strengths assessment tool] we got loads of criticism from colleagues in positive psychology asking how we could include weaknesses in a strengths tool. The answer is simple – our clients and organisations, and real people in the real world – were telling us that strengths are great, but I also need to know what is going to trip me up, and it needs to be balanced and realistic.' (provider of strengths-based workplace training, interview, July 2013)

The core messages of positive psychology – an emphasis on both the economic and societal value of optimism and self-optimisation – are in many ways an easy sell to business leaders and HR managers. Many practitioners of positive psychology approaches at work noted that this has been particularly the case for business facing austere times since the economic recession, where their approach:

> '... offers a really interesting promise, which is people will work the hardest not for money, not for reward, not for pay, not for punishment but for pleasure, for satisfaction. Now you've got to imagine how enormously appealing that is to a company.' (director of a business development company, interview, July 2013)

The science

Aside from the business rationale for positive psychology practice in the workplace and its influence on HR, management and organisational studies, positive psychology has rapidly evolved from a movement to an academic sub-discipline. While positive psychologists themselves often trace back the history of this research focus to those pioneering 1930s organisational studies, inter-war studies of giftedness, marital happiness, effective parenting and meaning in life (Seligman and Csikszentmihalyi, 2000), or the humanist psychologists of the 1950s and 1960s such as Abraham Maslow (Diener, 2009, p 7), the current enthusiasm for positive psychology has a much more recent history, dating back to Martin Seligman's presidency of the American Psychological Society in 1998. While Seligman and fellow US-based psychologists had spent their careers to date investigating concepts such as 'learned optimism' (Seligman, 2006 [1990]), 'hope' (Snyder, 2000), 'strengths' (through the work of Don Clifton and the Gallup organisation[4]) and 'flow' (Csikszentmihalyi, 1990), it was not until the late 1990s that concerted efforts were made to actively come together as a network of positive psychologists in order to address psychologists' 'half-baked' stories about the human mind (Seligman, 2006 [1990], p iii).

Initial meetings were held in Philadelphia, Washington DC, Lincoln, Lawrence, Columbia, Grand Cayman and Akumal (Lopez and Gallagher, 2009, p 3), and by the mid-2000s, positive psychology had established Master's programmes (in applied positive psychology [MAPP]) and research centres in the US, UK and Australia. Not only did these courses bring positive psychology into the heart of graduate academic training, but the MAPP qualification also produced a large number of organisational and HR consultants. By 2005, positive psychology featured in hundreds of undergraduate courses and nearly 30 major programmes existed at US universities (Yen, 2010, p 68). By 2011, several academic journals had been launched (*Journal of Happiness Studies* since 2000, *The Journal of Positive Psychology* since 2006 and *Psychology of Well-being* since 2011), and a number of influential textbooks had been published (Linley and Joseph, 2004; Peterson, 2006; Baumgardner and Crothers, 2009; Lopez and Snyder, 2009; Hefferon and Boniwell, 2011). Seligman was to bring his work to popular and policy attention with the publication of his best-selling texts, *Learned optimism* (2006 [1990]), *Authentic happiness* (2002) and *Flourish* (2011).

---

[4]   www.thestrengthsfoundation.org/

There was a certain amount of initial scepticism of the efforts to establish positive psychology as part of the academic mainstream, as one senior lecturer and positive psychology business consultant recalls: "I remember sometime in 2001/02 mentioning it in the BPS [British Psychological Society], the very senior levels [that I] do positive psychology and literally having a reaction that people laugh in my face" (interview, August 2013). In the face of such criticism, it is not uncommon for a textbook in positive psychology to begin with an assertion of its scientific credentials, or an assertion of what it is not (positive thinking, pseudo-science, 'mumbo-jumbo', self-help). There is still very much a sense that the 'case for' positive psychology is still being made (Lopez and Gallagher, 2009, p 3).

In light of the appeal that the scientific prowess of positive psychology appears to have among business, it is of interest that so many positive psychology publications draw attention to its scientific methods, paraphernalia, communities and rigour. Sarah Lewis, author of *Positive psychology at work* (2011, p 4), puts it as follows:

> ... [p]ositive psychology is further distinguished from positive thinking by the fact that it has "body of knowledge" structures such as collegiate bodies, university departments, professors and rigorous accredited academic courses ... it has all the paraphernalia of scientific discourse with peer-reviewed journals and academic conferences. Its practitioners apply to respected scientific bodies for research grants. Assertions made as fact can be checked, verified or refuted by others.

And yet, it is widely recognised by practitioners of positive psychology in business contexts that the general 'scientificness' of the approach is a crucial selling point for their work. The scientifically validated methods for creating positive workers and positive workplace cultures are a key part in distinguishing positive psychology from other 'fads' which feature regularly in the business and management world:

> 'Businesses are very faddish, they always like the latest thing, especially if it has come from academia and it sounds quite scientific, especially if it is to do with the brain and especially if they think it is something that will help the business perform better.' (provider of strengths-based workplace training, interview, July 2013)

The making of positive psychology as a science, as critical psychologist Jeffrey Yen has demonstrated, has relied on many practical and institutional advances such as the building of the aforementioned 'paraphernalia of scientific discourse' (for example, courses, books and journals, conferences, international associations such as the International Positive Psychology Association, and its 'Global Chinese', European and Canadian counterparts). Yen also notes just how well positive psychology has been financially resourced, through Seligman's successes in securing research funding and the pivotal role of the John Templeton Foundation which has funded research, prestigious and high-value financial prizes and research institutes and networks (2010). Supporter Diener (2009, p 8) judges that part of its success is due to 'Seligman's charisma and organizational skills'. His strategic approach to establishing a productive network of researchers has also been noted, as has the persuasive rhetorical style of writing in its major texts and seminal writings (Yen, 2010).

Yen's contemporary historicisation and critical reading of the discourse of positive psychology is one of the few of its kind to elaborate on the careful construction of this science as a solution to topical problems and as a socially valued form of knowledge, so we will revisit the detail of his critique in the final section of the chapter. In drawing attention to the 'making' of the science of positive psychology, my aim is not to question its credentials from a scientific basis. Rather, it is to demonstrate how science itself is historically made within particular geo-historical circumstances, and thus offers only a partial account of human activity. It is not the final grounds for truth, but an intersubjective agreement between people (professionally trained-scientists) as to the best estimation of the state of current knowledge. This is, of course, a basic account of the sociology of science. It is nonetheless helpful in investigating the way in which the knowledge of positive psychology is complexly intertwined with its application in practice in one of its key spaces of intervention: the workplace.

'Feeling good and functioning well' (optimism and optimisation) at work

> 'So I talk about, you know, feeling good and functioning well and that's quite a helpful discussion to have with businesses because it's about, you know, this isn't just about helping people feel better and feel happy all the time, it's also about how they function in the world and, you know.' (former business HR manager and current positive psychology consultant, interview, July 2013)

The combination of 'feeling good and functioning well' is an appealing promise provided by workplace training providers. And it is no coincidence that positive psychology has become so popular as an approach to organisational management and workplace training. This co-evolution is part of a more obvious economisation of positive emotion shared also with the enthusiasm for happiness economics in public policy. One of the main claims of positive psychology in workplace training is that it is based not simply on positive thinking or some other flaky notion, but on the *science of optimism*. The science of optimism is characterised by quite some debate over just what optimism is. Seligman's early animal research in the 1960s (which involved giving electric shocks to dogs when they attempted to escape an enclosed space), as recalled in his popular text, *Learned optimism*, had showed that optimism was a *skill* that could be learned rather than a relatively fixed personality trait.

Along with his colleague, Steve Maier, Seligman found that the dogs' behaviour was in fact determined by their *expectation* of receiving an electric shock; over time the dogs learned to be helpless and stopped even trying to escape their enclosures. This pessimism was later 'trained out' of them by the researchers, and their interventions (teaching the dogs that they could respond to the shocks) were said to give the dogs a positive outlook. Seligman later trialled his theories with workers in a life insurance business. Those who had been recruited for their optimism were found to deal with rejection better, pick themselves up and showed higher sales figures. Seligman's later work at his VIA Institute includes 'hope' as a *character strength* (a personality trait of moral value). At other times, he has described hope and optimism *as subjective experience* (Seligman and Csikszentmihalyi, 2000, p 5), and subsequently, as a *habit of thinking* or an *attitude* (Seligman, 2006 [1990], pp 4-5). Optimism as an attitude explains why some people see bad fortune or defeat as:

> ... just a temporary setback, that its causes are confined to this one case. The optimists believe defeat is not their fault: circumstances, bad luck, or other people brought it about.[...] Confronted by a bad situation, they perceive it as a challenge and try harder. (Seligman, 2006 [1990], p 4)

Both Seligman and his frequent co-author, Peterson, have also searched for the biological roots of optimism, finding Lionel Tiger's *Optimism: The biology of hope* (1979) to be a significant contribution to understanding the fundamental human nature and basic science of evolutionary optimism (Peterson, 2000, p 46; Seligman, 2006 [1990], p 108). Optimism for positive psychologists in the workplace is a 'state of explanation' (Lewis, 2011, p 152) rather than a state of mind. It refers to Seligman's account of how we explain the causes of events to ourselves, and is said to have positive benefits, not only in terms of workplace performance, but also with respect to our mental and physical health, our resilience and life satisfaction (Lewis, 2011, p 152). For psychologists of positive emotions such as Michael Cohn and Barbara Fredrickson (2009, p 14), it is this quality of *explanation* which makes something like optimism worth researching as an emotion with real effects. They have set out to provide scientific evidence of the real physical, physiological and psychological benefits of positive emotions through experimental studies. At the heart of their project is a concern to show the role of positive emotions in 'humanity's toolbox for growth' (Cohn and Fredrickson, 2009, p 21).

Scientists will not appreciate the somewhat inconsistent definitions of optimism characteristic of this literature, but whether optimism is a learned skill, a trait (strength), a state, an attitude or a style, what seems to be clear is that optimism is open to change, capable of growth, responsive to self-management, and amenable to optimisation. As the antithesis of 'learned helplessness' (of those dogs and life insurance salespeople), optimism is intimately associated with a sense of autonomy or personal control. Seligman and his colleagues' own work was indeed in part a challenge to the Pavlovian behaviourism of postwar psychology which still reduced learning to the model of stimulus–response, and evacuated personal autonomy from the equation. Instead, optimism as an explanatory style was now avowedly conceptualised as a cognitive potentiality, something to be worked on within representational schema rather than relegated to the behavioural world of automaticity (Peterson, 2000, p 46).

It is therefore only a short step from optimism to optimisation, the imperative clearly marked by positive psychology in the workplace to improve oneself, to use 'humanity's toolbox for growth' in the pursuit of, well, more growth. Positive psychology remains readily adaptable in terms of what forms such self-optimisation should take, of what exactly should be optimised. This is possible because it has already settled the 24 key characteristics of the timeless and universally morally valuable life summarised in the word 'flourishing' (Seligman, 2011). As such, it is the *normalisation* of optimisation which is the key process here. Seligman explains the rationale for optimisation as rooted in contemporary society. The development of character virtues and training of positive emotions will address the moral deficiencies left by the decline in religiosity, marriage and sociability. While people may be biologically limited in terms of their 'set range' of happiness, according to Seligman (2002), *authentic happiness* is under your voluntary control; it is educable, and it is therefore your responsibility to optimise.

In the deployment of positive psychology in the workplace, this imperative to optimise is a characteristic of a number of approaches, not least the AI method of change management. Based on the ideas of David Cooperrider, a business consultant, speaker and management scholar, AI regards organisations not as problems to be solved, but as 'high commitment work systems' in which a focus on optimising workers' commitments, passions, success stories and strengths can engender more organic organisational change (Cooperrider, 2007, p 22). For AI practitioners, optimism must be optimised in a business context of continuous change:

> '… it's continuous change, it's always there, and if you had that definition, then I think it really strengthens the need for any of the positive stuff, because the only way to keep moving is constant change, is to continuously build on what you do well.' (director of strength-based organisational development company providing solutions focus and AI, interview, July 2013)

In other words, it is now crucial to develop the psychological resilience to cope emotionally, and the character strengths to consistently out-perform yourself in an uncertain work environment. Indeed, as one managing director of a strengths-based workplace training provider recounted, her number one strength from which she defines a clear sense of purpose is a 'maximiser':

> '… my No 1 strength is maximiser, so of course I would want to do the best, I would never settle for anything that wasn't top, so even though I didn't want to do it, I was going to give it my best shot, does that make sense?' (managing director of a strengths-based workplace training provider, interview, August 2013)

By focusing its solutions, products and services on optimism and optimisation, positive psychology in workplace training has found a home in a business environment which must pursue savings in an austere economic environment at the same time as achieving growth. The apparently unlikely scenario in which a person's happiness and their workplace performance is not always directly linked to their increasing salary, but more with their everyday wellbeing, provides workplace training providers with a novel market in providing 'austerity happiness'. As one practitioner explained, "you can squeeze adequacy out of anybody, but you can only motivate people to success" (practitioner in positive psychology and adviser on happiness at work, interview, May 2013). Whether it is a focus on strengths, talents, appreciation, authentic leadership, psychological capital, the neurobiology of positive emotion or mindfulness practice for wellbeing, positive psychology as a science alongside the business requirements of HR and management consultants are united in their resolve to achieve growth and success *in any context*. And yet, the business rationales of optimism and optimisation as well as the scientific paraphernalia which surround them are highly culturally, economically and politically context-dependent, as the final section goes on to elaborate.

The adequately satisfied worker of the American dream

The use of positive psychology in workplace training programmes is not aimed simply at cultivating positive sentiment, wellbeing or life satisfaction, but is predicated on a sense of personal autonomy through which workers are given a sense of responsibility to 'self-actualise' or indeed, 'self-maximise'. In other words, positive psychology views optimism as at once a biologically universal character virtue and a skill to be learned, trained and optimised. In one sense, Seligman himself seems to appreciate the historical contingency of the alignment of psychology with an individualised society of choice and personal control:

> For the first time in history – because of technology and mass production and distribution, and for other reasons – large numbers of people are able to have a significant measure of choice and therefore personal control over their lives. Not the least of these choices concerns our own habits of thinking. By and large, people have welcomed that control. We belong to a society that grants to its individual members powers they have never had before, a society that takes individuals' pleasures and pains very seriously, that exalts the self and deems personal fulfilment a legitimate goal, an almost sacred right. (Seligman, 2006 [1990], p 10)

Yet at the same time as insinuating the social construction of goal-driven personal fulfilment, Seligman (2006 [1990], p 6) also portrays the individual choice to direct our habits of thinking as a *biological developmental pathway* from the 'utter helplessness' of infancy towards gaining personal control.

Within the space of positive psychology's brief history, several psychologists from a more interpretivist tradition have investigated the cultural specificity of positive psychology in its presentation of itself as both a hard empirical science embedded in the psychological (and more recently the neuro)sciences and a transformational movement fundamentally at odds with the overriding negativity of mainstream psychology. Psychologist Barbara Held, for instance, has argued that the positive psychology movement 'does not exist in a cultural vacuum: it faithfully reflects a dominant ethos of American culture, namely, the need to "think positive," to be happy, healthy and wise' (Held, 2004, p 12). Held questions the basis on which Seligman and Peterson's inventory of strengths and character virtues was derived, and challenges their assertion that the theories of positive psychology are scientific and descriptive, rather than moralistic and prescriptive. Quite simply, she points out that the 24 universal and pan-historical character strengths were derived from the partial selection of particular cultural and religious texts chosen on specific moral grounds (Held, 2004, p 20). She contests that virtues can never be scientifically determined because by their very nature they prescribe 'what should be, not what is' (Held, 2004, p 20). In other words, they are part of a process of *norm formation* and not scientific facts.

Similarly Yen (2010) traces the rhetorical self-presentation of the positive psychology movement as a specifically *fin de siècle* project. From 1998 to 2000, he identifies a movement which describes its own development within an American culture of peace, prosperity, economic and social stability, relative affluence and progress. He describes how positive psychology in this context narrates its own value and purpose in terms of turning both the attentions of the discipline of psychology, and of society more generally, to what makes a good life (Yen, 2010, p 71). By 2001, however, this narrative had changed substantially to a focus on moral and social decay (for example, the decline in religiosity, marriage and the pursuit of individualistic, materialistic and hedonistic pleasures), and a need for psychological resilience in a context of insecurity and threat. The turning point, as Yen demonstrates, is marked by 9/11 and the effects these tragic events had on American cultural identity and national security in 2001. With specific reference to these events, positive psychologists began to re-frame the social purpose of the movement as one which resolved to support individuals and society to persevere in the face of adversity. Positive psychology's *raison d'être* has thus changed in a short space of time, and appears to conterminously pull in different directions, oscillating between:

> ... starkly contrasting images of American society, at first of the United States as a great beacon of civilization and progress, then as a deeply self-centered, superficial, and materialistic culture, and finally as a morally irreproachable sovereignty under siege. (Yen, 2010, p 72)

In addition to its role in ensuring the resilience of the US population under threat, positive psychology becomes a reaction against individualism. It contrasts itself to the hedonistic and narcissistic ideals of self-help, and yet it is also an avowedly individualistic project, through its own ideals of self-actualisation and personal fulfilment. In this way, positive psychology's universalised account of the human pursuit of flourishing serves to radically de-contextualise that very human self around which the movement is based.

For scholars such as Yen, the turn of the new millennium provided the geopolitical and national ideological context in which positive psychology gained rapid popularity as a necessary and inevitable solution to the prevailing problems facing the US at that time. For others, the decontextualising tendency of positive psychology is even more fundamental. Its foundational identification of universal human strengths is hampered by its unacknowledged situatedness within a specifically modern, Western philosophical, political and social framework. Psychologists Christopher and Hickinbottom (2008), for instance, argue that positive psychology is culturally ethnocentric. It makes problematic assumptions about the natural basis of the human self and the pursuit of happiness. The collapse of the pre-modern social, political and religious order heralded by the Enlightenment and American and French Revolutions gave way, they argue – not inevitably, but as a result of geo-historical circumstance – to an interiorised reflexive self, able to pass personal judgement on the nature of the good life. This self would have been unrecognisable to pre-modern and ancient societies within which moral norms were discernable only on an external basis, such as in relation to the natural cosmology, the social order or transcendental God (Christopher and Hickinbottom, 2008, p 567).

Moreover, Christopher and Hickinbottom (2008, p 574) are able to demonstrate in detail just how culturally specific the core assumptions of positive psychology turn out to be. They do so by identifying substantive differences in the conception of base concepts such as the 'individual', 'life satisfaction', 'emotions', 'positivity' and 'negativity' between predominantly Western traditions and a multitude of non-secular and/or non-Western cultural contexts and accounts of mind (not least within Buddhism, Yoga, Taoism, Confucianism, Jewish, Christian and Muslim practice, Balinese, Indigenous American, Hindu Indian, Japanese, Chinese cultures, as well as more communitarian or collectivist Western psychological traditions denoted by social interest or *Gemeinschaftsgefühl*).

In essence, Christopher and Hickinbottom's critique of positive psychology is also a critique of the foundational assumptions of Western psychology, its claims to know and delimit what we understand by 'personhood' and its propensity to disregard the geo-historical (economic, social, cultural, political) conditions from which the concept of the person derives meaning. The specific trouble with positive psychology is that it quite clearly does not take the socio-political state of the world sufficiently seriously in its valorisation of life satisfaction. That said, Seligman has to some extent responded to some of the criticisms of his approach. He is by now well known for having changed his mind about the value of happiness and satisfaction in favour of the broader notions of wellbeing and flourishing, which he regards as going beyond measures of positive mood and feeling (Seligman, 2011, p 14). In dismissing mere 'happiology' as a thing that can be measured (for example, life satisfaction on a scale of 1-10) with wellbeing theory as a *construct*, Seligman precisely concedes that positive psychology's original conception of positive emotion had been limiting. For now, Seligman's theory of wellbeing includes the elements known by another acronym, PERMA (positive emotion, engagement, positive relationships, meaning and accomplishment).

In part this addresses the criticisms of the cultural imperialism of the individualistic notion of life satisfaction, which bore no relation to intersubjective or social sources of meaning. And yet, the pursuit of meaning here is given no context; it remains for the individual as a choosing subject to determine meaning as 'something that you believe is bigger than the self' (Seligman, 2011, p 17). Such developments in the self-transcending sensitivities of wellbeing are thus not likely to appease critics such as Christopher and Hickinbottom (2008, p 576), since meaning remains de-contextualised from its specific sources, and the 'ends' which are to be accomplished remain resolutely individualistic:

'what human beings, when free of coercion, choose to do for its own sake' (Seligman, 2011, p 20). The unacknowledged commitment to liberal individualism here is for many critics of positive psychology a crucial problem. The source if this problem stems from tensions inherent in positive psychology's stated commitment to be descriptive/morally neutral/scientific *and yet* applied/solutions-focused/meaningful. Thus, through positive psychology's endeavours to establish a foundational account of meaning, its 'successful' identification of the universal truths of the meaningful life, and its scientific objectification of the determining elements of wellbeing, it occludes the historical struggles through which the liberal freedom to be the dominant source *and* arbiter of meaning have come to pass in specific places, cultures and eras.

In relation to the application of positive psychology in workplace training programmes, the political-economic context of the contemporary neoliberal workplace is therefore crucial to understanding developments in HR and management discourse. The pursuit of happiness, an optimistic orientation and the endeavour to achieve optimal individual functioning proposed by the happiness agenda and positive psychology are all part of a broader political economic story. It is part of a set of trends by which management theories and HR professionals are mobilised in the governance of positively-minded and *adequately* satisfied workers. Their achievement of full satisfaction, according to both Keynesian economic theory and neoliberal market logics, would precipitate capitalism's downfall (through a crash in consumer demand), and can thus only ever be promised, not experienced (Davies, 2011, p 71). And yet, the pursuit of happiness itself indicates a certain lack of contentment, along with a peculiar distancing of the self from one's present state of experience and one's actual context.

It is by 'feeling good and functioning well', particularly in the workplace, that we can fulfil our own potential within a free market in which employment is increasingly precarious. Feeling good (optimism) and functioning well (optimisation) work together in order to set behavioural norms and shape workplace culture. Happiness is therefore said to lie in our sense of agency and freedom to escape our social situation through learned exercises and training programmes which only work if we put in the emotional labour to be adequately optimistic. Workplace training programmes have become an important space for the cultivation of the skills and attributes required to *govern the self*, through psychological self-mastery and emotional management. Such programmes also promote personal optimism, hope and resilience in the face of uncertainty, workplace stress and 'frantic times'. The *normalisation* of optimisation through both positive psychology and the affective neuroscience of organisations is not, however, indicative of the inevitable consequences of scientific progress. Rather, it signifies the onward march of a specific political-economic rationality in which the worker, as with the learner of the previous chapter, must internalise a 'growth mindset'. For critics, positive psychology and popular neuroscience alike are part of a frontier mythology which is culturally specific to the US (Ehrenreich, 2009; Thornton, 2011).

The brain serves as the new frontier, a biological force over which the individual will eventually secure mastery. This signifies a person who is in constant pursuit of their (American) dreams (Becker and Marecek, 2008). Yet as this wider analysis of the geo-historical contextual rationalities of positive psychology at work has shown, this optimal person who is happy and sees the meaning of (working life) is not a universalised figure but one who – through the rhetorical framing and scientific practices of brain culture – becomes alienated from their specific context, and thus disempowered from discerning and changing it.

SIX

Conclusion:
what is at stake in the brain world?

From the bio-social to the psycho-spatial

As the varied manifestations of brain culture explored in this book indicate, neuroscientific, psychological and behavioural research has become increasingly influential in a range of fields of public policy and practice. Both the application of such research and the more diffuse discursive influence of brain culture are evident in a number of arenas in which our actions, behaviours and subjectivities are under formation. Brain-based explanations are now common in Western popular cultural understandings of ourselves, others and society, as well as in the kinds of research and knowledge which are (economically and politically) valued. But human action, behaviour and subjectivity are situated in particular geo-historical contexts, conditioned by prevailing social, economic, cultural and political rationalities, and are shaped by and within the context of brain culture itself. The circularity of this brain world thus plays a constitutive role in each of the particular social contexts investigated throughout the book. I have sought to identify in each of these worlds what is at stake in terms of the governance of citizens. How does the re-framing of human 'nature' through brain culture have an impact on the assumptions that governments, experts and actors make about governing the self?

This book has explored how narratives concerning our predictable irrationality, error-prone decision-making and automatic, pre-conscious action are at work in re-imagining the citizen as a political agent, and as a target of governance. It has considered the specific techniques used in governing citizens through brain culture in different social spheres, and the practical processes of subjectification that are mobilised in brain-targeted interventions. Specific forms of scientific knowledge, research networks and academic infrastructures are called on to rationalise the goals of self-actualisation, self-mastery and self-optimisation. New experts have been called forth to direct our attention, orchestrate our experiences and shape our spaces of action in this brain world. In sum, new forms of behavioural governance have begun to imply a new relation to our (un)willing selves.

The ascendance of brain culture is not necessarily a revolution in thought and practice. It is part of a much longer history in which the role of the brain in determining human character, agency and social mores has been widely debated. Nor is it simply the case that scientific progress is shaping our culture in new ways – rather, that a specific Western culture has shaped scientific investigation, and continues to do so in new ways in particular contexts. Science is now produced 'in the context of application' (Nowotny et al, 2001, p 1), and this is evident in the three spheres outlined in this book. In the brain world of neuroarchitecture, the demands for micro-scale interventions in spatial situations for the optimisation of health and learning are informed by a political vision in which the individual citizen can be responsibilised, and their immediate and affective behaviour governed by experts who offer body and brain-based explanations for that behaviour. On a larger scale, neuroarchitecture offers to create vast biophysical datasets that could be used for managing the spatial behaviours of populations, shaping 'programmable cities' (Kitchen, 2011), and directing attention away from other ways of understanding the character and drivers of urban experience. However, it is not just that neuroarchitecture *could* have these unintended consequences, but that the pursuit of neuroarchitecture is fashioned *by* a context in which social governance has already been personalised. The 'Smart City' in which the parameters of biophysical habits can be known and logistically managed is thus posited as a vehicle for health, learning and optimal functioning in the quest for competitive economic advantage.

In the brain world of neuroeducation, brain-based teaching and learning programmes, products and services can have the effect of misinforming teachers about the real educational value of such programmes. This approach can also medically pathologise students who do not fit neatly with the norms of a performance-driven educational system in which schools and nations compete in a knowledge economy, and can reduce learning to an algorithm to be optimised. And yet, this is only possible because neuroeducation is pursued as a solution to educational problems in a context in which teachers have long been de-professionalised and learning instrumentalised. Children who do not fit within the culturally specific norms of reading, writing and even sitting still are already alienated in this world. Intransigent social problems are again re-framed through behavioural forms of governance in which the brain is targeted as both the problem and solution to educational difference, while the structural inequalities of a geographically uneven educational landscape are obscured from political view.

Finally, in the brain world of the workplace, positive psychology and organisational neuroscience can have the cumulative impact of internalising the pursuit of adequate satisfaction at work. In pursuing brain-based approaches to positivity and emotional management, the mind and body of the worker are mobilised towards organisational goals, profitability and success. Again, brain culture does not simply help organisations to achieve this, but the research sciences of positive psychology, positive organisational behaviour, authentic leadership development and organisational neuroscience are themselves shaped by a context in which happiness has already been economised and paid work privileged as the rightful source of personal identity, social value and self-worth. From a critical perspective, these practices signify the culturally specific American dream scientised – the brain becomes a frontier to be known and mastered. In this sense, emotional labour and affective comportment are mobilised in order to secure psychological resilience in the face of economic and national threat. Optimism and optimisation become survival strategies in this fast-paced world, and the worker is simply assumed to have personal autonomy.

In all three spheres, scientific endeavour is directed towards expanding knowledge of ourselves at the expense of knowledge of the world beyond our visual perceptions and affective responses to it. And yet, while knowledge, investigation and measurement of the brain world abounds, along with a proliferation of applications, strategies and insights to inform further research, policy and practice, the brain world poses a specific challenge for social science, arts and humanities research. Such research is well-placed to develop our understandings of the circularity of brain culture and our evolving relationship to the brain world.

The proliferation of the 'neuro' prefix has been welcomed within numerous academic disciplines, policy arenas and commercial ventures, which have embraced brain culture in part to secure their future relevance. As I have examined throughout the book, brain culture is more than a battle for evidence and explanation in an academic sense; it has real effects, unintended consequences and cumulative impacts. These are evident in a range of social spheres in which both discourses and schemes of action address their attention to the brain, psyche and behaviour. But brain culture also raises fundamental philosophical questions concerning determinism and free will, moral responsibility, political agency and ethical judgement – concerns shared as much by people in general as by professional philosophers. For some, this adds up to an essential re-thinking of human subjectivity – we are more animal, more biophysical, less rationalising and more functional and material than we thought we were. So, too, it is said, we should re-think the basis for our social interactions. These can now be accounted for in terms of the science of herd behaviours, our neurobiological internalisation of cultural norms, our neural capacity for empathy, our brains' plasticity, or our epigenetic susceptibility to external, social and environmental influence.

For critics, residual concerns over the often necessarily reductionist scientific method pursued in the psychological and behavioural sciences remain unresolved. Furthermore, the overblown claims of neuroscientific research (especially the various applications of such research findings and the apparently seductive allure of the visual 'evidence' of brain-scanning images as they circulate through popular culture) are said to pose a substantial threat to what we understand our human agency and personhood to be. There are concerns that this brain culture has the effect of 'aping' humankind and de-contextualising complex social phenomena. There is a sense that in the hasty search for the origins of consciousness, decision-making, emotion, reason, action and personhood within highly localised regions of the brain, we risk portraying a deterministic account of human action and medicalising social phenomena. In this brain world, it is argued, we are losing sight of what it means to be reflexive thinkers, as opposed to being irrational automatons who are biophysically determined, acting before we think.

Throughout the book I have sought to forge an approach that addresses both some of these philosophical concerns and the practical repercussions of brain culture as they relate to the governance of citizens in specific contexts. In each of the brain worlds discussed, evidence can be found of sources of criticism and reflection that attest to people's refusal to be defined as neural subjects. While the specific processes of subjectification outlined in Chapter One are operant, they are not necessarily successful. Indeed, there is much ordinary resistance to the potential reductionism, medicalising and determinist tendencies of the brain sciences in their practical application. In Chapter Three, for example, concerns were expressed within neuroarchitecture over the necessary reductionism of the field, and the likely political problems associated with 'scaling' up neuroarchitecture. In Chapter Four, Educational Psychologists (EPs) were vocal in their criticisms of neuroeducation's tendency to narrow attention on medicalised 'within child' explanations of educational success and failure. And in Chapter Five, psychologists expressed concerns that positive psychology's universalised account of biological developmental pathways obscured the cultural, political and economic imperatives to 'think positive' and 'self-maximise'. I have tried to show how the kind of reflexive thinking we do *about* the brain world is itself shaped *by* the brain world and its various neural turns. Our accounts of the brain world are thus contingent on the contextual rationalities that have shaped us as diverse people in specific geo-historical circumstances.

In tackling actually existing brain culture in this way, it is important to consider how we can most effectively analyse the political stakes that emerge in practising brain culture without over-claiming its novelty and force, over-stating some privileged position from which we might stage critique, or conjuring a sweeping rejection of science itself as a politically suspect force. At the same time, we need to appreciate the partiality of behavioural, psychological and neuroscientific knowledge amidst their apparent superiority as modes of explanation – as evidenced by the increasing uptake of such bodies of knowledge in numerous fields of research, policy and practice. One way of doing this is to explain more clearly the value of diverse (social, cultural, political, geographical, historical and literary and so on) understandings of human action, behaviour and subjectivity without clinging on to new advancements in science in search of a *raison d'être*. Another is to identify the partial and situated claims of those sciences that are posited as the basis for politically neutral forms of action and evidence-based policy.

From the outset, we must acknowledge that the sciences in general are already – in the contemporary UK and North American academic contexts at least – posited as institutionally and professionally superior to the arts, humanities and social sciences. This is clearly reflected in the disproportionate amount of funding, media coverage and status associated with the scientific disciplines. In 2011, the Campaign for Social Science was set up in the UK specifically to address a lack of political voice for the social sciences, their low media profile, issues with funding the publication of social science research, a decline in postgraduate study, and threats to existing longitudinal datasets used by social science researchers.[1] Funding for the political and social sciences is also deemed under threat in the US and Canada (Plazek and Steinberg, 2013), while the rapidly changing economic and political climate of contemporary higher education puts the value of the arts, humanities and social sciences at risk (Benneworth and Jongbloed, 2010). In this context, increased interdisciplinary has become routine. In specific relation to developments in the neurosciences and social theory, a

---

[1]   https://campaignforsocialscience.org.uk/

'bio-social' approach to research has been proposed (Harrington et al, 2006, p 4). Here I consider the prospects for this interdisciplinary bio-social science, and propose an alternative 'psycho-spatial' approach for shedding light on the governance of citizens within the brain world in practice.

Bio-social science

There have been long-running debates concerning the intersections and conflicts between scientific and sociological theory that have served to re-contextualise the scientific endeavour through pursuing varied forms of critique and interpretation. Notable in its engagement of this type is the school of thought termed Science and Technology Studies (STS), associated with sociologists such as Bruno Latour, Steve Woolgar, John Law and Anne-Marie Mol. From the 1970s, these authors, and many others since, have paid close attention to the everyday enactment of the scientific method, the 'production' of scientific facts, the role of technology in scientific research and application, and public understandings of science (see Law, 2008, for a detailed history of STS).

In undertaking anthropological and qualitative inquiry of and among scientists, STS situates scientific findings in their social and historical context. In many cases, STS seeks to expose whose interests have been served by the particular use of scientific technologies, and identifies the partial standpoints from which apparently objective facts have been constructed. In their seminal work, *Laboratory life*, Latour and Woolgar (1986 [1979], p 32, original emphasis) describe their concern 'with the *social* construction of scientific knowledge in so far as it draws attention to the *process* by which scientists make sense of their observations.' In observing everyday social interactions in scientific laboratories, they describe how scientists (and themselves as anthropological observers) are invested in the craft of rendering the chaos of scientific practice into systematic statements that could be said to be scientific. Scientists are, in their words, 'routinely confronted by a seething mass of alternative interpretations. Despite participants' well-observed reconstructions and rationalisations, actual scientific practice entails the confrontation and negotiation of utter confusion' (Latour and Woolgar, 1986 [1979], p 36).

I recall this area of work because it signifies an important undercurrent to contemporary proposals for a bio-social science that have shaped academic debates on how best to scrutinise neuroscientific developments and brain culture. This is particularly the case within critical neuroscience, which draws heavily on the STS approach as a basis for critique (Slaby and Choudhury, 2012, p 42). The tenor of this debate is also evident within the discipline of human geography, which, as Chapter Two explored, has been struggling with its relationship with and/or status as a science for several decades. It has dealt with this struggle in various ways, and the recent 'neural turn' at work in non-representational geographies is one way in which this has been manifest. But for some, STS had gone too far in reporting the social construction of scientific facts and by implication, physical reality.

The physicist Alan Sokal infamously decried the excesses of social constructivism (albeit by lumping science studies together with a diverse swathe of postmodern cultural theory), suggesting that the humanities in general had lost their grip on social reality. He proclaimed: 'anyone who believes that the laws of physics are mere social conventions is invited to try transgressing those conventions from the windows of my apartment. I live on the twenty-first floor' (Sokal, 1996). Sokal published a spoof article in the cultural studies journal *Social Text* in order to expose the illogical and navel-gazing pretentions of what he characterised as postmodern theory. In what came to be known as the 'Sokal Affair', a political conflict within intellectual and academic culture was exposed between *science* and *interpretation* as expressions of truth (Guillory, 2002, p 479). Significant antagonisms between the sciences, the social sciences and cultural studies emerged in their claims to know and understand the human world (Guillory, 2002, p 482; see also Papoulias and Callard, 2010, p 32). When understood in the context of the inferior and declining status of the 'humanities', it is not therefore surprising that social constructionism and its associated interpretative methodologies have been devalued as a way of understanding the constitution of human nature.

High disciplinary stakes are evident, and these stakes go beyond 'mere' academic wrangling when the sciences are afforded an explanatory superiority concerning (1) all manner of questions of political agency and resistance, creative action and social change, and (2) programmes and policies which inform actual practice in a range of social worlds such as urban design, education and work. Some are optimistic about a potential rapprochement specifically between the life sciences and the social sciences. Nikolas Rose is a key figure in this endeavour, having co-founded a new journal, *BioSocieties*, in 2006 with historian of science Anne Harrington and bio-ethicist Ilina Singh.

The journal set out to encourage social theorists to take seriously transformations in the life sciences (including not only new scientific findings themselves, but also the regulation, application, technologisation and governance of scientific research and practice). It invited contributors to investigate how such developments are transforming society at a range of scales. Such investigations are a necessary intervention in the bio-ethical field that has been pre-occupied with bolting on 'ethical, legal and social implications' (consigned to the abbreviation 'ELSI') as an afterthought to the 'real' science. *BioSocieties* and the bio-social approach, they argue (Harrington et al, 2006, p 4), need to overcome the long-reaching divisions and disciplinary grand-standing which has been in evidence, between scientists who 'feel that the social sciences don't "believe in reality" or turn everything into a "social construction"', and by social scientists who are guilty of the 'recycling of critical nostrums about the socio-political role of biology.'

Extending this analysis of the revolutionary power of the life sciences, Rose (2013a, p 3) argues for a new kind of intellectual and institutional set of relations between the social and life sciences in order to go beyond description and critique, and to address a perceived crisis in legitimacy for the social sciences in a 'biological age'. Much progress has already been made. From the perspective of the social sciences, Rose surmises that 'it seems that "constructivism" is passé, the linguistic turn has reached a dead end and a rhetoric of materiality is almost obligatory' (2013a, p 4). Social theory has apparently embraced the biological re-framing of what it means to be an embodied human, and despite what Rose identifies as the tendency of the 'affect' and non-representational theorists (considered in Chapter Two) to adopt only the most populist versions of insights from the biological sciences which date rapidly, this embrace is generally to be welcomed. For Rose, it is no longer legitimate or useful to critique the biological sciences on the grounds of determinism, medicalisation and reductionism (Rose, 2013a, p 23). He identifies at least three significant transformations that render these traditional social critiques irrelevant. However, Rose seems to waver – perhaps inevitably in bridging disciplinary divides – between this optimistic embrace of the new life sciences and a deep suspicion of its consequences.

First, the rise of epigenomics (the study of gene–environment interactions) means that biological conceptions of living organisms are not reductive, but dynamic, emergent and open to influence from what he terms their 'milieu'. But on the other hand, Rose *is* critical of a new molecular style of thought which reduces human life to the 'material properties of cellular components' to be governed through new technological channels for managing digital data (2013a, p 5). Second, biology has become closely intertwined with intervention and 'technologisation', such that the very study of biology is often heavily indebted to forms of engineering at the molecular level (such as Dolly the sheep). Hence, biology is not determinist but opportunist – nothing is beyond biological manipulation. But still, Rose advises precaution on these 'fantasies of [biological] omnipotence' that have spawned a 'global bioeconomy' based on the manipulation and capitalisation of biological knowledge which serve particular political interests (Rose, 2013a, p 6). Third, the life sciences can no longer be considered separate from society, as they are becoming increasingly socially significant in terms of how we understand ourselves as people – in Novas and Rose's terms, we are becoming 'somatic' individuals (2000, cited in Rose, 2013a, p 7).

We see ourselves through a biomedical lens, we engage in techniques of self-management founded on a sense by which we can know and govern our bodies and minds in order to become the best possible version of our selves. This is most evidently the case in the processes of normalisation and subjectification outlined in the cases of neuroeducation and workplace training programmes offered in Chapters Four and Five. It is in this last sense that the approach of this book clearly resonates with Rose's bio-social approach, in recognising the constitutive nature of the brain world in shaping social practice and identity. Yet I do not share his optimism that the social and the biological or neuroscientific share an equal footing – the rhetorical power of the latter as an explanation of social phenomena and rationale for policy and practice remains dominant.

It is in a spirit of optimism and dialogue that Rose and colleagues spent the five years from 2007-12 enabling interdisciplinary discussion on the ethical, social and legal implications of the neurosciences through the European Neuroscience and Society Network (ENSN),[2] funded by the European Science Foundation. This network aimed to go beyond the predominantly American field of neuroethics to consider the ethical, political and social implications of neuroscience from an empirical, rather than abstract, perspective. The methods and concepts of the social sciences are used to examine current and recent advancements in neuroscience, in particular in relation to the politics of public health, neuroeconomies and neurochemical selfhood and difference (Rose, 2013b, p 4).

---

[2] www.kcl.ac.uk/sspp/departments/sshm/research/ENSN/European-Neuroscience-and-Society-Network.aspx

However, while Rose argues that the social sciences should not feel threatened by brain-based explanations of human affairs (Rose and Abi-Rached, 2013, p 3), others have more strongly argued that interdisciplinary dialogue may not be the best way to proceed. Some are concerned that such dialogue can serve to bolster claims of neuroscientific novelty and social transformation, implicating sociological discourse in the very creation of brain culture (Pickersgill, 2013, p 333). There are also concerns that Rose's desired rapprochement between the biological and social sciences over-writes some quite legitimate concerns about biological essentialism, reductionism and individualisation which remain a barrier to meaningful dialogue between the social and neurosciences (Cromby, 2004). Social psychologist John Cromby notes several problems that may impede interdisciplinary conversations.

First, there has to be a desire to engage from both 'sides', and Cromby finds little evidence of this on the side of the neurosciences. The endeavour of 'social neuroscience' is found to be particularly deterministic in its search for the neural drivers of social behaviours, and as such is rather closed to substantive sociological ideas (Cromby, 2007, p 152). Second, there are important issues with language, particularly in unsettled neuroscientific and social scientific conceptions of the 'self', which may be fluid, vary in different cultural contexts and defy comparison across different historical epochs. Third, for neuroscientists, brain states and processes determine behaviour. But for social scientists, humans act purposefully and creatively within pre-existing rules, standards and social norms through which behaviour is guided, rendered meaningful or 'proper'.

There is also a tendency common in the neurosciences to equate the brain with the person. Yet it is not the brain that thinks, believes or acts, but the person, and so to locate such agentic properties within the brain 'simply pushes the problem of mind-body dualism "inside" the brain … making social science contributions appear less relevant' (Cromby, 2007, p 161). Related to these conceptual issues are a set of methodological problems that may, for Cromby, be the hardest to overcome. The core techniques of fMRI scanning, psychological questionnaires or experimental methods are subject to methodological individualism – which takes anything 'social' to be a mere aggregate of individual action, decision and preference (Cromby, 2007, p 163). Not only does fMRI scanning put the research subject in the very odd social situation of being largely immobile inside a scanner, but the way in which social groups are accounted for in the very recruitment of participants is problematic. Their histories, biographies and distinguishing cultural, social, political and economic features are stripped down/controlled for/regarded as largely insignificant. In sum, there is much work to be done, linguistically, conceptually and methodologically, if the neurosciences are to avoid de-contextualising human subjectivity and to provide fuller, scaled-up explanations for actual human action and behaviour in specific situations in the world.

Others have argued for the need to reject the presumed value of interdisciplinarity altogether. Fitzgerald and Callard (2015), for instance, propose a more experimental approach in which the sociocultural and neurobiological are not seen as different (and unequal) forms of explanation, but as part of the same world. They suggest that the only way to move beyond (reductionist sociocultural) critique or conversely, the over-enthusiastic embrace of the neurosciences, is to engage in 'experimental entanglements' in which social scientists are embedded in neuroscientific research laboratories and involved in the production of such knowledge (Fitzgerald and Callard, 2015, p 3). Their approach posits that there is no socio-cultural sphere *beyond* the neurosciences from which to offer a privileged critical perspective; we are all implicated in the brain world now. While they are by no means ignorant of the hierarchical (and historically contingent) institutional divisions between the social sciences and neurosciences (Fitzgerald and Callard, 2015, p 19), their approach is, like that of Rose, optimistic towards the possibility of overcoming these divides. Yet this political reality underpins the current application of neuroscientific, behavioural and psychological knowledge in practice beyond the laboratory.

As this book has explored within the worlds of architecture, education and the workplace, the privileging of brain-based explanations has real effects for governance and citizenship. It connotes pre-scripted urban arrangements of idealised space, produces a new cadre of behavioural experts, and shapes specific norms and subjectivities according to a partial account of human behaviour largely taken out of context. It therefore remains paramount to better understand the context in which bio-social forms of knowledge achieve such status, to outline the specific processes by which such forms of knowledge shape people, and to indicate the various ways in which these processes are contested *in situ*.

The apparent incompatibility of a sociologically minded analysis of the radical uncertainties of human behaviour with a neurobiological imperative to pinpoint the human condition seems to be a significant sticking point standing in the way of a bio-social science. This book has considered, with reference to several academic disciplines, how this apparent impasse has centred on the relative importance of the biophysicality of the brain as a body part. This is contrasted with the sociocultural, philosophical and political-economic dynamics of people's bodies as a source of identity. If these dynamics are significant, which I think they are, then this body/identity is socially constituted in a specific geo-historical context. That context, as I have argued, is the case in the fields of practice outlined in the book, is the context of the brain world. While it is not new, the brain world is especially evident at the turn of the 21st century, particularly in the UK and US contexts studied here. This world plays a constitutive role in shaping research agendas, public policies and everyday practice. Handling the circularity of this phenomenon is a difficult issue, both philosophically and methodologically, and it is at the interface between the brain and world – in the psyche properly situated in its spatial context – that further exploration may be fruitful.

Psycho-spatial analysis

It is perhaps the position of the critical neuroscientists that provides the most confident assertion of the value of the social science traditions for analysing brain culture. They, too, are open to productive two-way dialogue that engages directly with neuroscientists in ways that avoid bolstering the revolutionary exhalations of 'neurotalk' (Choudhury and Slaby, 2012, p 2). They outline several bases for constructive interdisciplinary collaborations in which social and neuroscientists work together on: deliberating and refining parameters, concepts and categories used in laboratory experiments; investigating social and cultural phenomena as experienced by diverse social groups rather than the universal neural subject; 'enriching behavioural theories' by providing 'layered explanations of complex phenomena'; and investigating the discursive resonances or 'feedback loops' between popular cultural accounts of the brain and neuroscientific research (brain culture) (Slaby and Choudhury, 2012, p 43). But it is their treatment of situated subjectivity and of context that potentially offers the most productive basis for an empirical and sociological analysis of actually existing brain culture.

Attention to what we might term the 'psycho-spatial' is illuminating in acknowledging the vastly differing political and economic status of these contrasting insights into human behaviour. The term 'psycho-spatial' denotes that there is still much value in social scientists investigating the *mental* sphere as distinct from the neural, as a meaningful interface between brain and world. The psyche necessarily relates to the embodied mind in particular spaces at particular times, in particular fields of practice in context. This embodiment is not limited to affective or neurobiological explanation, but confers a socially situated and discursively enacted sense of personhood.

Like Choudhury and Slaby (2012), I find the nascent bio-social analyses to be too politically timid in their mapping of scientific practice and institutionalised forms of knowledge. By contrast, their critical neuroscience is arguably more sensitive to inequities of power, is more interested in scaled-up sociological explanations of neuroscientific practice, and has the potential to more accurately elucidate the processes of subject formation invoked by such practice. Their analytical project promises to investigate the dynamics of power that have rendered the contemporary neurosciences such a significant explanatory force. Of course, in order to offer such an analysis, which promises to 'penetrate beneath the surface' (Slaby and Choudhury, 2012, p 37), they have to hold on, to a certain extent, to a binary vision of the biological and the social. Critical neuroscience aims, after all, to 'de-naturalise' behavioural and individual phenomena as described by neuroscientists.

They set out to put neuroscience back into context, and to identify how its interpretive schema is sustained and reproduced in the service of particular social and political interests. They regard neuroscience itself as both socially constitutive and socially constituted, and thus seek to explore its contextual rationalities:

> Contextualizing neuroscientific objects of inquiry – whether the "neural basis" of addiction, depression, sociality, lying or adolescent behaviours – can, in this way, demonstrate how such findings, whilst capturing an aspect of behaviour in the world, are also held in place by a number of factors co-produced by a collection of circumstances, social interests, and institutions. (Slaby and Choudhury, 2012, p 31)

A focus on the 'psycho-spatial' similarly revives some unfashionable dualisms, but it could be argued that it is not the dualisms per se that are problematic, but their hierarchical opposition. A psycho-spatial analysis of brain culture is directed towards understanding the contextual rationalities by which particular and partial ways of knowing the world come to inform research, policy and practice, modes of governance and citizen identities. It offers a challenge to the re-framing of social problems as having brain-based solutions, holds on to the social as a scale of explanation for human behaviour, and pays attention to the geo-historical as a context for subjectivity. It sees the explanatory space between the psyche and the brain as a source of political agency from which to ask overtly democratic questions concerning contemporary manifestations of brain culture.

This approach pays attention to diverse ways of knowing the social, and not just the biological; of understanding the mind of the person, and not only the brain; of explaining the subjective 'me' as well as the agentic 'I'; and of accounting for geo-historical context beyond immanent experience and affect. A psycho-spatial approach does not take human behaviour to be determined by a convergence of environmental, neural and genetic material phenomena. Rather, it sees the human psyche as enacted in specific spatial contexts, and as a socio-spatial form of consciousness in which the boundaries of the psyche exceed the biological brain. The specific context that has long been in development is a brain world that remains to be fully understood.

References

Allen, G. and Duncan Smith, I. (2008) *Early intervention: Good parents, great kids, better citizens*, London: Centre for Social Justice and the Smith Institute.

Allin, P. (2007) 'Measuring societal wellbeing', *Economic & Labour Market Review*, vol 1, no 10.

Amin, A. and Thrift, N. (2013) *Arts of the political: New openings for the Left*, Durham, NC: Duke University Press.

Anderson, B. (2012) 'Affect and biopower: towards a politics of life', *Transactions of the Institute of British Geographers*, vol 37, pp 28-43.

Anderson, J. (2012) 'Stress in the city: a pioneering study explores the relationship between urban planning and human happiness', *Britain in 2013*, p 13.

Atkinson, S. (2011) 'Moves to measure wellbeing must support a social model of health', *British Medical Journal*, vol 343, no 7832, p 7323.

Atkinson, S., Fuller S. and Painter, J. (2012) 'Wellbeing and place', in S. Atkinson, S. Fuller and J. Painter (eds) *Wellbeing and place*, Farnham: Ashgate, pp 1-14.

Avey, J.B., Avolio, B.J., Crossley, C.D. and Luthans, F. (2009) 'Psychological ownership: theoretical extensions, measurement and relation to work outcomes', *Journal of Organizational Behavior*, vol 30, no 2, pp 173-91.

Avolio, B.J. and Gardener, W.L. (2005) 'Authentic leadership development: getting to the root of positive forms of leadership', *The Leadership Quarterly*, vol 16, pp 315-38.

Barker, R.G. and Wright, H.F. (1951) *One boy's day*, New York: Harper & Row.

Barkley, R.A. (2014) 'Sluggish cognitive tempo (concentration deficit disorder?): current status, future directions, and a plea to change the name', *Journal of Abnormal Child Psychology*, vol 42, no 1, pp 117-25.

Barnett, C. (2001) 'Culture, geography, and the arts of government', *Environment and Planning D: Society and Space*, vol 19, no 1, pp 7-24.

Barnett, C. (2008) 'Political affects in public space: normative blind-spots in non-representational ontologies', *Transactions of the Institute of British Geographers*, vol 33, no 2, pp 186-200.

Barnett, C. (2013) 'Book review essay: theory as political technology. Clive Barnett on Amin and Thrift's "Arts of the political"', *AntipodeFoundation.org/book-reviews* (http://wp.me/p16RPC-JO).

Barnett, C., Cloke, P., Clarke, N. and Malpass, A. (2005) 'Consuming ethics: articulating the subjects and spaces of ethical consumption', *Antipode*, vol 37, no 1, pp 23-45.

Barrett, P., Zhang, Y., Moffat, J. and Kobbacy K. (2013) 'A holistic, multi-level analysis identifying the impact of classroom design on pupils' learning', *Building and Environment*, vol 59, pp 678-89.

Barsade, S.G., Brief, A.C. and Spataro, S.E. (2003) 'The affective revolution in organizational behaviour. The emergence of a paradigm', in J. Greenburg (ed) *Organizational behaviour. The state of the science* (2nd edn), Mahwah, NJ: Lawrence Erlbaum Associates Publishers, pp 3-52.

Battro, A.M., Fischer, K.W. and Léna, P.J. (eds) (2008) *The educated brain: Essays in neuroeducation*, Cambridge: Cambridge University Press.

Baumgardner, S.R. and Crothers, M.K. (2009) *Positive psychology*, Upper Saddle River, NJ: Prentice Hall/Pearson Education.

Becker, D. and Marecek, J. (2008) 'Dreaming the American dream: individualism and positive psychology', *Social and Personality Psychology Compass*, vol 2, no 5, pp 1767-80.

Becker, W.J., Cropanzano, R. and Sanfey, A.G. (2011) 'Organizational neuroscience: taking organizational theory inside the neural black box', *Journal of Management*, vol 37, no 4, pp 933-61.

Benedek, M., Jauk, E., Fink, A., Koschutnig, K., Reishofer, G., Ebner, F. and Neubauer, A.C. (2014) 'To create or to recall? Neural mechanisms underlying the generation of creative new ideas', *NeuroImage*, vol 88, pp 125-33.

Benneworth, P. and Jongbloed, B.W. (2010) 'Who matters to universities? A stakeholder perspective on humanities, arts and social sciences valorisation', *Higher Education*, vol 59, no 5, pp 567-88.

Binkley, S. (2011a) 'Psychological life as enterprise: social practice and the government of neo-liberal interiority', *History of the Human Sciences*, vol 24, no 3, pp 83-102.

Binkley, S. (2011b) 'Happiness, positive psychology and the program of neoliberal governmentality', *Subjectivity*, vol 4, no 4, pp 371-94.

Blackwell, L.S., Trzesniewski, K.H. and Dweck, C.S. (2007) 'Implicit theories of intelligence predict achievement across an adolescent transition: a longitudinal study and an intervention', *Child Development*, vol 78, no 1, pp 246-63.

Blakemore, S.J. and Choudhury, S. (2006) 'Development of the adolescent brain: implications for executive function and social cognition', *Journal of Child Psychology and Psychiatry*, vol 47, no 3, pp 296-312.

Blakemore, S.-J. and Frith, U. (2000) *The implications of recent developments in neuroscience for research on teaching and learning, Report to the Teaching and Learning Research Programme* (www.tlrp.org/acadpub/Blakemore2000.pdf).

Blakemore, S.J. and Frith, U. (2005) *The learning brain. Lessons for education*, Oxford: Blackwell.

Brett, M., Johnsrude, I.S. and Owen, A.M. (2002) 'The problem of functional localization in the human brain', *Nature Reviews Neuroscience*, vol 3, no 3, pp 243-9.

Bruer, J. (1997) 'Education and the brain: a bridge too far', *Educational Researcher*, vol 26, no 8, pp 4-16.

Bruer, J. (2013) 'Afterword', in D. Mareschal, B. Butterworth and A. Tolmie (eds) *Educational neuroscience*, Chichester: John Wiley & Sons, pp 349-63.

Bunting, T.E. and Guelke, L. (1979) 'Behavioural and perception geography: a critical appraisal', *Annals of the Association of American Geographers*, vol 69, pp 448-62.

Burgdor, J. and Panksepp, J. (2006) 'The neurobiology of positive emotions', *Neuroscience and Biobehavioural Reviews*, vol 30, no 2, pp 173-87.

Butler, M.J.R. and Senior, C. (2007) 'Research possibilities for organizational cognitive neuroscience', *Annals of the New York Academy of Sciences*, vol 1118, pp 206-10.

Buttimer, A. (1971) *Society and milieu in the French geographic tradition*, Chicago, IL: Rand McNally.

Cacioppo, J.T. and Berntson, G.G. (1992) 'Social psychological contributions to the decade of the brain: doctrine of multilevel analysis', *American Psychologist*, vol 47, no 8, pp 1019-28.

Cadman, L. (2010) 'How (not) to be governed: Foucault, critique, and the political', *Environment and Planning D: Society and Space*, vol 28, no 3, pp 539-56.

Caragliu, A., Del Bo, C.F. and Nijkamp, P. (2009) *Smart cities in Europe*, No 48, Serie Research Memoranda, Amsterdam: Faculty of Economics, Business Administration and Econometrics, VU University.

Caspi, A. and Moffitt, T.E. (2006) 'Opinion: gene–environment interactions in psychiatry: joining forces with neuroscience', *Nature Reviews Neuroscience*, vol 7, pp 583-90.

Chiao, J.Y. (2009) 'Cultural neuroscience: a once and future discipline', *Progress in Brain Research*, vol 178, pp 287-304.

Choudhury, S. and Slaby, J. (2012) 'Introduction. Critical neuroscience – between lifeworld and laboratory', in S. Choudhury and J. Slaby (eds) *Critical neuroscience. A handbook of the social and cultural contexts of neuroscience*, Chichester: Wiley-Blackwell, pp 1-26.

Choudhury, S., Fishman, J.R., McGowan, M.L. and Juengst, E.T. (2014) 'Big data, open science and the brain: lessons learned from genomics', *Frontiers in Human Neuroscience*, vol 8, no 239, pp 1-10.

Christopher, J.C. and Hickinbottom, S. (2008) 'Positive psychology, ethnocentrism, and the disguised ideology of individualism', *Theory and Psychology*, vol 18, no 5, pp 563-89.

Clark, G.L. (2011) 'Myopia and the global financial crisis: short-termism, context-specific reasoning, market structure and institutional governance', *Dialogues in Human Geography*, vol 1, pp 4-25.

Clark, J. (2013) 'Philosophy, neuroscience and education', *Educational Philosophy and Theory*, vol 47, no 1, pp 36-46 (http://dx.doi.org/10.1080/00131857.2013.866532).

Cohen, D. (2006) 'Critiques of the "ADHD" enterprise', in G. Lloyd, J. Stead and D. Cohen (eds) *Critical new perspectives on ADHD*, Abingdon: Routledge, pp 12-33.

Cohn, M.A. and Fredrickson, B.L. (2009) 'Positive emotions', in S.J. Lopez and C.R. Snyder (eds) *The Oxford handbook of positive psychology* (2nd edn), Oxford: Oxford University Press, pp 13-24.

Cohn, S. (2011) 'Visualizing disgust', in F. Ortega and F. Vidal (eds) *Neurocultures. Glimpses into an expanding universe*, Frankfurt am Mein: Peter Lang, pp 181-98.

Colls, R. and Evans, B. (2009) 'Measuring fatness, governing bodies: the spatialities of the Body Mass Index (BMI) in anti-obesity politics', *Antipode*, vol 41, no 5, pp 1051-83.

Cooper, Z. and Fairburn, C.G. (2001) 'A new cognitive behavioural approach to the treatment of obesity', *Behaviour Research and Therapy*, vol 39, no 5, pp 499-511.

Cooperrider, D. (2007) 'Business as an agent of world benefit. Going green @ maximum velocity', *Global HR News*, p 22.

Cox, K.R. (1981) 'Bourgeois thought and the behavioural geography debate', in K.R. Cox and R.G. Golledge (eds) *Behavioural problems in geography revisited*, London: Methuen, pp 256-80.

Cox, K.R. and Golledge, R.G. (eds) (1981) *Behavioural problems in geography revisited*, London: Methuen.

Crockett, M.J. (2009) 'The neurochemistry of fairness. Clarifying the link between serotonin and prosocial behavior', *Annals of the New York Academy of Sciences*, vol 1167, pp 76-86.

Crogan, P. and Kinsley, S. (2012) 'Paying attention: towards a critique of the attention economy', *Culture Machine*, vol 13, pp 1-29 (www.culturemachine.net).

Cromby, J. (2004) 'Between constructionism and neuroscience. The societal co-constitution of embodied subjectivity', *Theory & Psychology*, vol 14, no 6, pp 797-821.

Cromby, J. (2007) 'Integrating social science with neuroscience: potentials and problems', *BioSocieties*, vol 2, pp 149-69.

Cromby, J. (2011) 'The greatest gift? Happiness, governance and psychology', *Social & Personality Psychology Compass*, vol 5, no 11, pp 840-52.

Cromby, J. and Willis, M.E.H. (2014) 'Nudging into subjectification: governmentality and psychometrics', *Critical Social Policy*, vol 34, no 2, pp 241-59.

Csikszentmihalyi, M. (1990) *Flow*, New York: Harper & Row.

Damasio, A.R. (1994) *Descartes' error: Emotion, reason and the human brain*, London: Picador.

Da Silva, A.N.R., Zeile, P., Aguiar, F.d.O., Papastefanou, G. and Bergner, B.S. (2014) 'Smart sensoring and barrier free planning: project outcomes and recent developments', in N.N. Pinto, J.A. Tenedório, A.P. Antunes and J.R. Cladera (eds) *Technologies for urban and spatial planning: Virtual cities and territories*, Hershey, PA: IGI Global, pp 93-112.

Davidson, R.J. and Sutton, S.K. (1995) 'Affective neuroscience: the emergence of a discipline', *Current Opinion in Neurobiology*, vol 5, pp 214-17.

Davies, W. (2011) 'The political economy of unhappiness', *New Left Review*, vol 71 (Sept-Oct), pp 65-80.

Davies, G. and Beech, A. (eds) (2012) *Forensic Psychology: Crime, justice, law, interventions*, Chichester: John Wiley & Sons Ltd.

Della Sala, S. and Anderson, M. (eds) (2012) *Neuroscience in education. The good, the bad and the ugly*, Oxford: Oxford University Press.

Devine-Wright, P. and Clayton, S. (2010) 'Introduction to the special issue: place, identity and environmental behaviour', *Journal of Environmental Psychology*, vol 30, no 3, pp 267-70.

de Vos, J. (2014) 'The death and the resurrection of (psy)critique. The case of neuroeducation', *Foundations of Science*, pp 1-17 (http://link.springer.com/article/10.1007%2Fs10699-014-9369-8).

DfES (UK Department for Education and Skills) (2003) *Building schools for the future*, London: DfES.

DfES (2006) *2020 vision: The report of the Teaching and Learning in 2020 Review Group*, Nottingham: DfES.

Dickinson, Z.C. (1933) 'Review of workers' emotions in shop and home by Rexford B. Hersey', *Journal of Political Economy*, vol 41, no 3, pp 425-27

Diener, E. (2009) 'Positive psychology: past, present and future', in S.J. Lopez and C.R. Snyder (eds) *The Oxford handbook of positive psychology* (2nd edn), Oxford: Oxford University Press, pp 7-11.

Doidge, N. (2007) *The brain that changes itself. Stories of personal triumph from the frontiers of brain science*, London: Penguin.

Dolan, P., Hallsworth, M., Halpern, D., King, D. and Vlaev, I. (2010) *Mindspace. Influencing behaviour through public policy*, London: Institute for Government and Cabinet Office.

Durante, K.M. and Saad, G. (2010) 'Ovulatory shifts in women's social motives and behaviors: implications for corporate organizations', in A.A. Stanton, M. Day and I.M. Welpe (eds) *Neuroeconomics and the firm*, Cheltenham: Edward Elgar, pp 116-30.

Eberhard, J.P. (2006a) 'Foreword', in J. Zeisel, *Inquiry by design. Environment/behaviour/neuroscience in architecture. Interiors, landscape, and planning* (2nd edn), London: W.W. Norton & Company, pp 11-12.

Eberhard, J.P. (2006b) 'Hypotheses as examples of PhD thesis projects', Unpublished document.

Eberhard, J.P. (2009) *Brain landscape. The co-existence of neuroscience and architecture*, Oxford: Oxford University Press.

Ecclestone, K. (2012) 'From emotional and psychological well-being to character education: challenging policy discourses of behavioural science and "vulnerability"', *Research Papers in Education*, vol 27, no 4, pp 463-80.

Edelstein, E. (2014) 'Translating neuro-architecture from cell to city', Paper presented at the ESRC Behaviour Change and Psychological Governance Seminar Series, Bristol, 24 March.

Edelstein, E.A. Gramann, K., Schulze, J., Bigdely, N., Elke van Erp, S., Vankov, A. Makeig, S., Wolszon, L. and Macagno, E. (2008) 'Neural responses during navigation in the virtual aided design laboratory: brain dynamics of orientation in architecturally ambiguous space', in S. Haq, C. Hölscher and S. Torgrude (eds) *Movement and orientation in built environments: Evaluating design rationale and user cognition*, SFB/TR 8 Report No 015-05/2008, Bremen: Universität Bremen/Universität Freiburg, Bremen (www.sfbtr8.uni-bremen.de/en/home/).

Edwards, R., Gillies, V. and Horsley, N. (2013) 'Rescuing Billy Elliot's brain: neuroscience and early intervention', Paper presented at 'Brain Science and Early Intervention', Joint meeting of the BSA Childhood Study Group and the BSA Families and Relationships Study Group, Goldsmiths University, London, 20 June.

Ehrenberg, A. (2011) 'The "social" brain: an epistemological chimera and a sociological truth', in F. Ortega and F. Vidal (eds) *Neurocultures. Glimpses into an expanding universe*, Frankfurt am Main: Peter Lang, pp 117-40.

Ehrenreich, B. (2009) *Smile or die. How positive thinking fooled America and the world*, London: Granta Publications.

Elden, S. (2001) *Mapping the present: Heidegger, Foucault and the project of a spatial history*, New York: Continuum.

Fine, C. (2010) *Delusions of gender. How our minds, society, and neurosexism create difference*, London: W.W. Norton.

Fitzgerald, D. and Callard, F. (2015) 'Social science and neuroscience beyond interdisciplinarity: experimental entanglements', *Theory, Culture and Society*, vol 32, no 1, pp 3-32.

Foucault, M. (2003) '*Society must be defended*': Lectures at the Collège de France, 1975-1976, New York: Picador.

Foucault, M. (2007) *Security, territory, population*, Translated by Graham Burchell, New York: Palgrave MacMillan.

Fox, E. (2008) *Emotion science: Cognitive and neuroscientific approaches to understanding human emotions, Basingstoke:* Palgrave Macmillan.

Fredrickson, B. (2001) 'The role of positive emotions in positive psychology: the broaden-and-build theory of positive emotions', *American Psychologist*, vol 56, pp 218-26.

Frosh, S. (2001) 'Psychoanalysis, identity and citizenship', in N. Stevenson (ed) *Culture and citizenship*, London: Sage Publications, pp 62-73.

Furedi, F. (2011) *On tolerance: A defence of moral independence*, London: Continuum.

Gage, F. (2009) 'Foreword. From the Perspective of a Neuroscientist', in P. Eberhard, *Brain landscape. The coexistence of neuroscience and architecture*, Oxford: Oxford University Press, ppxii-xiv.

Gagen, E. (2013) 'Governing emotions: citizenship, neuroscience and the education of youth', *Transactions of the Institute of British Geography* (http://onlinelibrary.wiley.com/doi/10.1111/tran.12048/pdf).

Gardiner, E. (2012) *Changing behaviour by design. Combining behavioural science with design-thinking to help organisations tackle big social issues*, London and Coventry: Design Council and Warwick Business School.

Geake, J. (2008) 'Neuromythologies in education', *Educational Research*, vol 50, no 2, pp 123-33.

Geiger, B.M., Haburcak, B., Avena, N.M., Moyer, M.C., Hoebel, B.G. and Pothosa, E.N. (2009) 'Deficits of mesolimbic dopamine neurotransmission in rat dietary obesity', *Neuroscience*, vol 159, no 4, pp 1193-9.

George, B. (2003) *Authentic leadership*, San Francisco, CA: Jossey-Bass.

Gibbs, D., Krueger, R. and MacLeod, G. (2013) 'Grappling with smart city politics in an era of market triumphalism', *Urban Studies*, vol 50, no 11, pp 2151-7.

Gold, J. (1980) *An introduction to behavioural geography*, Oxford: Oxford University Press.

Goldacre, B. (2006) 'Brain Gym exercises do pupils no favours', *The Guardian*, 18 March, p 13.

Golledge, R.G. and Stimson, R.J. (1987) *Analytical behavioural geography*, London: Croom Helm.

Goswami, U. (2006) 'Neuroscience and education: from research to practice?', *Nature Reviews Neuroscience*, vol 7, pp 406-13.

Goswami, U. (2008) *Cognitive development. The learning brain*, Hove: Psychology Press.

Graham, L.J. (2010) 'Teaching ADHD?', in L.J. Graham (ed) *(De) constructing ADHD. Critical guidance for teachers and teacher educators*, New York: Peter Lang Publishing, Inc, pp 1-19.

Greco, M. and Stenner, P. (2013) 'Happiness and the art of life: diagnosing the psychopolitics of wellbeing', *Health, Culture and Society*, vol 5, no 1, pp 1-19.

Grist, M. (2009) *Changing the subject. How new ways of thinking about human behaviour might change politics, policy and practice*, London: RSA.

Grist, M. (2010) *Steer. Mastering our behaviour through instinct, environment and reason*, London: RSA.

Guillory, J. (2002) 'The Sokal Affair and the history of criticism', *Critical Inquiry*, vol 28, pp 470-508.

Häkli, J. and Kallio, K.P. (2013) 'Subject, action and polis: theorizing political agency', *Progress in Human Geography*, vol 38, no 2, pp 181-200.

Hannah, M. (2000) *Governmentality and the mastery of territory in nineteenth-century America*, Cambridge: Cambridge University Press.

Hannah, M. (2013) 'Attention and the phenomenological politics of landscape', *Geografiska Annaler: Series B, Human Geography*, vol 95, no 3, pp 235-50.

Hannah, S.T., Woolfolk, R.L. and Lord, R.G. (2009) 'Leader self-structure: a framework for positive leadership', *Journal of Organizational Behavior*, vol 30, no 2, pp 269-90.

Hardiman, M.M. (2012) *The brain-targeted teaching model for 21st-century schools*, London: Corwin.

Hardiman, M.M., Rinne, L., Gregory, E. and Yarmolinskaya, J. (2012) 'Neuroethics, neuroeducation, and classroom teaching: where the brain sciences meet pedagogy', *Neuroethics*, vol 5, no 2, pp 135-43.

Harrington, A., Rose, N. and Singh, I. (2006) 'Editors' introduction', *BioSocieties*, vol 1, pp 1-5.

Harvey, D. (1989) *The urban experience*, Oxford: Basil Blackwell.

Hefferon, K. and Boniwell, I. (2011) *Positive psychology. Theory, research and applications*, Maidenhead: Open University Press.

Held, B.S. (2004) 'The negative side of positive psychology', *Journal of Humanistic Psychology*, vol 44, no 1, pp 9-46.

Helliwell, J., Layard, J. and Sachs, R. (eds) (2012) *World happiness report*, New York: The Earth Institute, Columbia University (www.earth. columbia.edu/articles/view/2960).

Herrick, C. (2009) 'Designing the fit city: public health, active lives, and the (re)instrumentalization of urban space', *Environment and Planning* A, vol 41, no 10, pp 2437-54.

Hibbing, J. (2013) 'Ten misconceptions concerning neurobiology and politics', *Perspectives on Politics*, vol 11, no 2, pp 475-89.

Hochschild, A. (2003 [1983]) *The managed heart. Commercialization of human feeling*, London: University of California Press.

Hogan, P. (2014) 'Literary brains: neuroscience, criticism, and theory', *Literature Compass*, vol 11, no 4, pp 293-304.

Hollands, G.J., Shemilt, I., Marteau, T.M., Jebb, S.A., Kelly, M.P., Nakamura, R., Suhrcke, M. and Ogilvie, D. (2013) 'Altering micro-environments to change population health behaviour: towards an evidence base for choice architecture interventions', *BMC Public Health*, vol 13, p 1218.

Hollands, R.G. (2008) 'Will the real smart city please stand up? Intelligent, progressive or entrepreneurial?', *City*, vol 12, no 3, pp 303-20.

House of Lords (2011) *Behaviour Change, Science and Technology Select Committee*, 2nd report of session 2010-12, HL Paper 179, London: The Stationery Office.

Howard-Jones, P. (2007) *Neuroscience and education: Issues and opportunities. A commentary by the Teaching and Learning Research Programme (TLRP)*, London: Teaching and Learning Research Programme.

Howard-Jones, P. (2010) *Introducing neuroeducational research: Neuroscience, education and the brain from contexts to practice*, Abingdon: Routledge.

Huppert, F.A. and Baylis, N. (2004) 'Well-being: towards an integration of psychology, neurobiology and social science', *Philosophical Transactions of the Royal Society B: Biological Sciences*, vol 359, no 1449, pp 1447-51.

Huppert, F.A. and So, T.C.C. (2011) 'Flourishing across Europe: application of a new conceptual framework for defining well-being', *Social Indicators Research*, vol 110, no 3, pp 837-61.

Hurley, S. (1998) *Consciousness in action*, Cambridge, MA: Harvard University Press.

Hurley, S. and Noë, A. (2003) 'Neural plasticity and consciousness', *Biology and Philosophy*, vol 18, pp 131-68.

Huxley, M. (2006) 'Spatial rationalities: order, environment, evolution and government', *Social and Cultural Geography*, vol 7, no 5, pp 771-87.

Isen, A.M. (2009) 'A role for neuropsychology in understanding the facilitating influence of positive affect on social behaviour and cognitive processes', in S.J. Lopez and C.R. Snyder (eds) *The Oxford handbook of positive psychology* (2nd edn), Oxford: Oxford University Press, pp 503-18.

Isin, E.F. (2004) 'The neurotic citizen', *Citizenship Studies*, vol 8, no 3, pp 217-35.

Isin, E.F. (2009) 'Citizenship in flux: the figure of the activist citizen', *Subjectivity*, vol 29, pp 367-88.

Jahnukainen, M. (2010) 'Different children in different countries: ADHD in Canada and Finland', in L.J. Graham (ed) *(De)constructing ADHD. Critical guidance for teachers and teacher educators*, New York: Peter Lang Publishing, Inc, pp 63-76.

James, O. (2007) *Affluenza*, London: Random House.

Jensen, E. (2007) *Introduction to brain-compatible learning* (2nd edn), London: Sage Publications.

JISC (Joint Information Systems Committee) (2006) *Designing spaces for effective learning. A guide to 21st century learning space design*, Bristol: Higher Education Funding Council for England.

Jones, R., Pykett, J. and Whitehead, M. (2011a) 'The geographies of soft paternalism in the UK: the rise of the avuncular state and changing behaviour after neoliberalism', *Geography Compass*, vol 5, no 1, pp 50-62.

Jones, R., Pykett, J. and Whitehead, M. (2011b) 'Governing temptation: on the rise of libertarian paternalism in the UK', *Progress in Human Geography*, vol 35, no 4, pp 483-501.

Jones, R., Pykett, J. and Whitehead, M. (2013) *Changing behaviours. On the rise of the psychological state*, Cheltenham: Edward Elgar.

Kabat-Zinn, J. (2004 [1994]) *Wherever you go, there you are. Mindfulness meditation for everyday life*, London: Piaktus.

Kabat-Zinn, J. (2011) 'Foreword', in M. Williams and D. Penman (eds) *Mindfulness: A practical guide to finding peace in a frantic world*, London: Piatkus, pp ix–xii.

Kean, B. (2009) 'ADHD in Australia: the emergence of globalization', in S. Timimi and J. Leo (eds) *Rethinking ADHD. From brain to culture*, Basingstoke: Palgrave Macmillan, pp 169–97.

Kitchin, R. (2011) 'The programmable city', *Environment and Planning B: Planning and Design*, vol 38, no 6, pp 945–51.

Kitchen, R. and Blades, M. (2002) *The cognition of geographic space*, London: I.B. Tauris & Co Ltd.

Kluemper, D.H., Little, L.M. and DeGroot, T. (2009) 'State or trait: effects of state optimism on job-related outcomes', *Journal of Organizational Behavior*, vol 30, no 2, pp 209–31.

Korf, B. (2008) 'A neural turn? On the ontology of the geographical subject', *Environment and Planning A*, vol 40, no 3, pp 715–32.

Krabbendam, L. (2005) 'Schizophrenia and urbanicity: a major environmental influence – conditional on genetic risk', *Schizophrenia Bulletin, vol* 31, no 4, pp 795–9.

Kraftl, P. (2012) 'Utopian promise or burdensome responsibility? A critical analysis of the UK government's building schools for the future policy', *Antipode*, vol 44, no 3, pp 847–70.

Kraftl, P. (2013) *Geographies of alternative education. Diverse learning spaces for children and young people*, Bristol: Policy Press.

Kraftl, P. (2014) 'Liveability and urban architectures: mol(ecul)ar biopower and the "becoming lively" of sustainable communities', *Environment and Planning D: Society and Space*, vol 32, no 2, pp 274–92.

Krause, S.R. (2008) *Civil passions: Moral sentiment and democratic deliberation*, Oxford: Princeton University Press.

Larner, W. (2000) 'Neo-liberalism: policy, ideology, governmentality', *Studies in Political Economy*, vol 63, pp 5–25.

Latour, B. and Woolgar, S. (1986 [1979]) *Laboratory Life. The construction of scientific facts*, Princeton, NJ: Princeton University Press.

Law, J. (2008) 'On sociology and STS', *Sociological Review*, vol 56, no 4, pp 623–49.

Layard, R. (2005) *Happiness: Lessons from a new science*, London: Allen Lane.

Lefebvre, H. (1991 [1974]) *The production of space*, Translated by D. Nicholson-Smith, Oxford: Blackwell.

Lewis, S. (2011) *Positive psychology at work. How positive leadership and appreciative inquiry create inspiring organizations*, Chichester: Wiley-Blackwell.

Ley, D. (1981) 'Behavioural geographies and the philosophies of meaning', in K.R. Cox and R.G. Golledge (eds) *Behavioural problems in geography revisited*, London: Methuen, pp 209-30.

Linley, A. and Joseph, S. (eds) (2004) *Positive psychology in practice*, Hoboken, NJ: John Wiley & Sons.

Linley, A., Harrington, S. and Garcea, N. (eds) (2010) *Oxford handbook of positive psychology and work*, Oxford: Oxford University Press.

Lopez, S.J. and Gallagher, M.W. (2009) 'A case for positive psychology', in S.J. Lopez and C.R. Snyder (eds) *The Oxford handbook of positive psychology* (2nd edn), Oxford: Oxford University Press, pp 3-6.

Lopez, S.J. and Snyder, C.R. (eds) (2009) *The Oxford handbook of positive psychology* (2nd edn), Oxford: Oxford University Press.

Lowenthal, D. (1961) 'Geography, experience, and imagination: towards a geographical epistemology', *Annals of the Association of American Geographers*, vol 51, no 3, pp 241-60.

Luthans, F. and Youssef, C.M. (2009) 'Positive workplaces', in S.J. Lopez and C.R. Snyder (eds) *The Oxford handbook of positive psychology* (2nd edn), Oxford: Oxford University Press, pp 579-88.

Luthans, F., Avey, J.B., Avolio, B.J., Norman, S.M. and Combs, G.J. (2006) 'Psychological capital development: toward a micro-intervention', *Journal of Organizational Behavior*, vol 27, no 3, pp 387-93.

Lynch, K. (1960) *The image of the city*, London: MIT Press.

MacKerron, G. and Mourato, S. (2013) 'Happiness is greater in natural environments', *Global Environmental Change*, vol 23, no 5, pp 992-1000.

MacKuen, M., Marcus, G.E., Neuman, W.R. and Keele, L. (2007) 'The third way: the theory of affective intelligence and American democracy', in G.E. Marcus, W.R. Neuman, M. MacKuen and A. Crigler (eds) *The affect effect. Dynamics of political thinking and behaviour*, London: University of Chicago Press, pp 124-51.

McCann, E.J. (2003) 'Framing space and time in the city: urban policy and the politics of spatial and temporal scale', *Journal of Urban Affairs*, vol 25, no 2, pp 159-78.

Malik, S. (2013) 'Jobseekers' psychometric test "is a failure"', *The Guardian*, 6 May.

Mareschal, D., Butterworth, B. and Tolmie, A. (eds) (2013) *Educational neuroscience*, Chichester: Wiley-Blackwell.

Marteau, T.M., Hollands, G.J. and Fletcher, P.C. (2012) 'Changing human behaviour to prevent disease: the importance of targeting automatic processes', *Science*, vol 337, no 6101, pp 1492-5.

Massey, D. (1975) 'Behavioural research', *Area*, vol 7, pp 201-3.

Masten, A.S. (2001) 'Ordinary magic: resilience process in development', *American Psychologist*, vol 56, no 3, pp 227-39.

Miller, P. and Rose, N. (1988) 'The Tavistock Programme: the government of subjectivity and social life', *Sociology*, vol 22, no 2, pp 171-92.

Moodie, R., Stuckler, D., Monteiro, C., Sherond, N., Neal, B., Thamarangsi, T., Lincoln, P., Casswell, S. on behalf of The Lancet NCD Action Group (2013) 'Profits and pandemics: prevention of harmful effects of tobacco, alcohol, and ultra-processed food and drink industries', *The Lancet*, vol 381, no 9867, pp 670-9.

Morton, J. and Frith, U. (1995) 'Causal modelling: a structural approach to developmental psychopathology', in D. Cicchetti and D.J. Cohen (eds) *Manual of developmental psychopathology*, New York: Wiley, pp 357-90.

Nathan, P.E. (2009) 'Foreword', in S.J. Lopez and C.R. Snyder (eds) *The Oxford handbook of positive psychology* (2nd edn), Oxford: Oxford University Press, pp xxiii-xxvii.

Neuman, W.R., Marcus, G.E., Crigler, A.N. and MacKuen, M. (2007) 'Theorizing affect's effects', in G.E. Marcus, W.R. Neuman, M. MacKuen and A. Crigler (eds) *The affect effect. Dynamics of political thinking and behaviour*, London: University of Chicago Press, pp 1-20.

Noë, A. (2009) *Out of our heads. Why you are not your brain, and other lessons from the biology of consciousness*, New York: Hill & Wang.

Nold, C. (2009) *Emotional cartography. Technologies of the self* (www.emotionalcartography.net).

Nowotny, H., Scott, P. and Gibbons, M. (2001) *Re-thinking science: Knowledge and the public in an age of uncertainty*, Cambridge: Polity Press.

Nussbaum, M. (2001) *Upheavals of thought: The intelligence of emotions*, Cambridge: Cambridge University Press.

O'Connor, C., Rees, G. and Joffe, H. (2012) 'Neuroscience in the public sphere', *Neuron*, vol 74, no 2, pp 220-6.

OECD (Organisation for Economic Co-operation and Development) (2007) *Understanding the brain: The birth of a learning science*, Paris: OECD (www.oecd.org/edu/ceri/understandingthebrainthebirthofalearningscience.htm).

Ortega, F. (2009) 'The cerebral subject and the challenge of neurodiversity', *Biosocieties*, vol 4, no 4, pp 425-45.

Osborne, T. and Rose, N. (1999) 'Do the social sciences create phenomena? The example of public opinion research', *British Journal of Sociology*, vol 50, no 3, pp 367-96.

Oullier, O. (2012) 'Le cerveau et la loi: éthique et pratique du neurodroit', Centre d'analyse stratégique (www.strategie.gouv.fr).

Oullier, O. and Sauneron, S. (2010) *Improving public health prevention with behavioural, cognitive and neuroscience*, Paris: Centre d'analyse stratégique.

Panksepp, J. (1998) *Affective neuroscience: The foundations of human and animal emotions*, Oxford: Oxford University Press.

Papoulias, C. and Callard, F. (2010) 'Biology's gift: interrogating the turn to affect', *Body and Society*, vol 16, no 1, pp 29-56.

Pedersen, C.B. and Mortensen P.B. (2001) 'Evidence of a dose–response relationship between urbanicity during upbringing and schizophrenia risk', *Archives of General Psychiatry*, vol 58, no 11, pp 1039-46.

Peterson, C. (2000) 'The future of optimism', *American Psychologist*, vol 55, no 1, pp 44-55.

Peterson, C. (2006) *A primer in positive psychology*, New York: Oxford University Press.

Peterson, C. and Park, N. (2009) 'Classifying and measuring strengths of character', in S.J. Lopez and C.R. Snyder (eds), *The Oxford handbook of positive psychology* (2nd edn), Oxford: Oxford University Press, pp 25-33.

Peterson, C. and Seligman, M.E.P. (2004) *Character strengths and virtues: A handbook and classification*, New York and Washington, DC: Oxford University Press and American Psychological Association.

Peterson, C., Park, N., Hall, N. and Seligman, M.E.P. (2009) 'Zest and work', *Journal of Organizational Behavior*, vol 30, no 2, pp 161-72.

Philo, C. (1992) 'Foucault's geography', *Environment and Planning D: Society and Space*, vol 10, pp 137-61.

Philo, C. (2012) 'A "new Foucault" with lively implications – or "the crawfish advances sideways"', *Transactions of the Institute of British Geographers*, vol 37, no 4, pp 496-514.

Pickering, S.J. and Howard-Jones, P. (2007) 'Educators' views on the role of neuroscience in education: findings from a study of UK and international perspectives', *Mind, Brain, and Education*, vol 1, no 3, pp 109-13.

Pickersgill, M. (2013) 'The social life of the brain: neuroscience in society', *Current Sociology*, vol 61, no 3, pp 322-40.

Pickersgill, M., Cunningham-Burley, S. and Martin, P. (2011) 'Constituting neurologic subjects: neuroscience, subjectivity and the mundane significance of the brain', *Subjectivity*, vol 4, pp 346-65.

Pile, S. (1996) *The body and the city. Psychoanalysis, space and subjectivity*, London: Routledge.

Pile, S. (2010) 'Emotions and affect in recent human geography', *Transactions of the Institute of British Geographers*, vol 35, no 1, pp 5-20.

Plazek, D.J. and Steinberg, A. (2013) 'Political science funding black out in North America? Trends in funding *should not* be ignored', *Political Science & Politics*, vol 46, no 3, pp 599-604.

Prokhovnik, R. (1999) *Rational woman: A feminist critique of dichotomy*, London: Routledge.

Protevi, J. (2009) *Political affect: Connecting the social and the somatic*, Minneapolis, MN: University of Minnesota Press.

Randerson, J. (2008) 'Experts dismiss educational claims of Brain Gym programme', *The Guardian,* 3 April (www.theguardian.com/science/2008/apr/03/brain.gym).

Randolph, J.J. (2013) 'Promoting psychosocial and cognitive wellness in the workplace: the emerging neuroscience of leadership development', in J.J. Randolph (ed) *Positive neuropsychology. Evidence-based perspectives on promoting cognitive health*, New York: Springer, pp 103-19.

Richardson, A. (2001) *British romanticism and the science of the mind*, Cambridge: Cambridge University Press.

Resch, B. (2013) 'People as sensors and collective sensing – contextual observations complementing geo-sensor network measurements', in J. Krisp (ed) *Advances in location-based service*, Berlin and Heidelberg: Springer, pp 391-406.

Robinson, E.L. and Higgs, S. (2013). 'Food intake norms increase and decrease snack food intake in a remote confederate study', *Appetite*, vol 65, pp 20-4.

Rose, H. (2004) 'Consciousness and the limits of neurobiology', in D. Rees and S. Rose (eds) *The new brain sciences. Perils and prospects*, Cambridge: Cambridge University Press, pp 59-70.

Rose, N. (1989) *Governing the soul: The shaping of private life* (2nd edn), London: Routledge.

Rose, N. (2003) 'Neurochemical selves', *Society*, vol 41, no 1, pp 46-59.

Rose, N. (2007) 'Governing the will in a neurochemical age', in S. Maasen and B. Sutter (eds) *On willing selves: Neoliberal politics vis-à-vis the neuroscientific challenge*, Basingstoke: Palgrave Macmillan, pp 81-99.

Rose, N. (2013a) 'The human sciences in a biological age', *Theory, Culture and Society*, vol 30, no 1, pp 3-34.

Rose, N. (2013b) *European neuroscience and society network. Final report*, London: King's College London (www.kcl.ac.uk/sspp/departments/sshm/research/ensn/European-Neuroscience-and-Society-Network.aspx).

Rose, N. and Abi-Rached, J.M. (2013) *Neuro. The new brain sciences and the management of the mind*, Woodstock: Princeton University Press.

Rowson, J. (2011) *Transforming behaviour change: Beyond nudge and neuromania*, London: RSA.

Royal Society, The (2011a) *Brain waves module 4: Neuroscience and the law*, London: The Royal Society.

Royal Society, The (2011b) *Brain waves module 2: Neuroscience: Implications for education and lifelong learning*, London: The Royal Society.

Rozin, P., Scott, S., Dingley, M., Urbanek, J.K., Jiang, H. and Kaltenbach, M. (2011) 'Nudge to nobesity I: minor changes in accessibility decrease food intake', *Judgment and Decision Making*, vol 6, no 4, pp 323-32.

Saint-Paul, G. (2011) *The tyranny of utility. Behavioural social science and the rise of paternalism*, Princeton, NJ: Princeton University Press.

Sample, I. (2014) 'Arguments over brain simulation come to a head', *The Guardian*, 7 July, p 5.

Satel, S. and Lilienfeld, S.O. (2013) *Brainwashed: The seductive appeal of mindless neuroscience*, New York: Basic Books.

Seligman, M. (2002) *Authentic happiness: Using the new positive psychology to realize your potential for lasting fulfilment*, New York: Free Press.

Seligman, M. (2006 [1990]) *Learned optimism. How to change your mind and your life*, New York: Vintage Books.

Seligman, M. (2011) *Flourish. A new understanding of happiness and well-being – and how to achieve them*, London: Nicholas Brealey.

Seligman, M. and Csikszentmihalyi, M. (2000) 'Positive psychology. An introduction', *American Psychologist*, vol 55, no 1, pp 5-14.

Selten, J. and Cantor-Graae, E. (2007) 'Hypothesis: social defeat is a risk factor for schizophrenia?', *The British Journal of Psychiatry*, vol 191, no 51, s9-s12.

Seymour, B. and Vlaev, I. (2012) 'Can, and should, behavioural neuroscience influence public policy?', *Trends in Cognitive Sciences*, vol 16, no 9, pp 449-51.

Shonkoff, J.P. and Phillips, D.A. (eds) (2000) *From neurons to neighborhoods: The science of early childhood development*, Washington, DC: The National Academies Press.

Simmons, B.L., Gooty, J., Nelson, D.L. and Little, L.M. (2009) 'Secure attachment: implications for hope, trust, burnout, and performance', *Journal of Organizational Behavior*, vol 30, no 2, pp 233-47.

Simon, H.A. (1947) *Administrative behavior: A study of decision-making processes in administrative organization*, New York: Macmillan.

Singer, W. (2008) 'Epigenesis and brain plasticity in education', in A.M. Battro, K.W. Fischer and P.J. Léna (eds) *The educated brain: Essays in neuroeducation*, Cambridge: Cambridge University Press, pp 97-109.

Slaby, J. (2010) 'Steps towards a critical neuroscience', *Phenomenology and the Cognitive Sciences*, vol 9, no 3, pp 397-416.

Slaby, J. and Choudhury, S. (2012) 'Proposal for a critical neuroscience' in S. Choudhury and J. Slaby (eds) *Critical neuroscience. A handbook of the social and cultural contexts of neuroscience*, Chichester: Wiley-Blackwell, p 29-51.

Smith, G., Jongeling, B., Hartmann, P., Russell, C. and Ladau, L. (2010) *Raine ADHD study: Long term outcomes associated with stimulant medication in the treatment of ADHD in children*, Department of Health, Government of Western Australia.

Smith, L. (2014) 'John Bruer: "Growing up in poverty doesn't damage your brain irretrievably"', *The Guardian*, 7 May.

Snyder, C.R. (2000) *Handbook of hope*, San Diego, CA: Academic Press.

Snyder, C.R., Sympson, S.C., Ybasco, F.C., Borders, T.F., Babayak, M.A. and Higgins, R.L. (1996) 'Development and validation of the State Hope Scale', *Journal of Personality and Social Psychology*, vol 70, no 2, pp 321-35.

Sokal, A. (1996) 'A physicist experiments with cultural studies', *Lingua Franca*, vol 6, May.

Sommers, S. (2011) *Situations matter. Understanding how context transforms your world*, London: Penguin.

Soon, C.S., Brass, M., Heinze, H.-J. and Haynes, J.-D. (2008) 'Unconscious determinants of free decisions in the human brain', *Nature Neuroscience*, vol 11, pp 543-5.

Sousa, D. (ed) (2010) *Mind, brain and education: Neuroscience implications for the classroom*, Bloomington, IN: Solution Tree Press.

Steg, L., van den Berg, A.E. and de Groot, J.I.M. (2012) *Environmental psychology: An introduction*, Hoboken, NJ: Wiley-Blackwell.

Stenner, P. and Taylor, D. (2008) 'Psychosocial welfare: reflections on an emerging field', *Critical Social Policy*, vol 28, no 4, pp 415-37.

Sternberg, E. (2009) *Healing spaces: The science of place and well-being*, London: Belknap Press.

Stiglitz, J.E., Sen, A. and Fitoussi, J. (2009) *Report by the Commission on the Measurement of Economic Performance and Social Progress* (www.stiglitz-sen-fitoussi.fr).

Stiles, A. (2012) *Popular fiction and brain science in the late nineteenth century*, Cambridge: Cambridge University Press.

Strauss, K. (2008) 'Re-engaging with rationality in economic geography: behavioural approaches and the importance of context in decision-making', *Journal of Economic Geography*, vol 8, no 2, pp 137-56.

Tallis, R. (2011) *Aping mankind. Neuromania, Darwinitis and the misrepresentation of humanity*, Durham: Acumen.

Taylor, D. (2011) 'Wellbeing and welfare: a psychosocial analysis of being well and doing well enough', *Journal of Social Policy*, vol 40, no 4, pp 777-94.

Taylor, K. (2012) *The brain supremacy: Notes from the frontiers of neuroscience*, Oxford: Oxford University Press.

Taylor, M. (2007) 'Changing minds: preparing for an era of neurological reflexivity' (www.thersa.org/events/video/archive/matthew-taylor).

Taylor, M. (2011) 'The century of the brain?', 15 November (www.matthewtaylorsblog.com/socialbrain2/the-century-of-the-brain/).

Thaler, R.H. and Sunstein, C. (2008) *Nudge. Improving decisions about health, wealth and happiness*, London: Penguin.

Thien, D. (2005) 'After or beyond feeling? A consideration of affect and emotion in geography', *Area*, vol 37, no 4, pp 450-6.

Thornton, D.J. (2011) *Brain culture: Neuroscience and popular media*, Piscataway, NJ: Rutgers University Press.

Thrift, N. (2004) 'Intensities of feeling: towards a spatial politics of affect', *Geografiska Annaler*, vol 86B, pp 57-78.

Thrift, N. (2007) *Non-representational theory: Space, politics, affect*, London: Routledge.

Thrift, N. (2008) 'I just don't know what got into me: where is the subject?', *Subjectivity*, vol 22, pp 82-9.

Tiger, L. (1979) *Optimism: The biology of hope*, New York: Simon & Schuster.

Timimi, S. and Leo, J. (2009) 'Introduction', in S. Timimi and J. Leo (eds) *Rethinking ADHD. From brain to culture*, Basingstoke: Palgrave Macmillan, pp 1-17.

Tokuhama-Espinosa, T. (2008) 'The scientifically substantiated art of teaching: a study in the development of standards in the new academic field of neuroeducation (mind, brain, and education science)', PhD thesis, Capella University.

Tokuhama-Espinosa, T. (2011) *Mind, brain, and education science: A comprehensive guide to the new brain-based teaching*, London: W.W. Norton & Co.

Tolia-Kelly, D.P. (2006) 'Affect – an ethnocentric encounter? Exploring the "universalist" imperative of emotional/affectual geographies', *Area*, vol 38, pp 213-17.

Tuan, Y.-F. (1971) 'Geography, phenomenology, and the study of human nature', *The Canadian Geographer*, vol 15, no 3, pp 181-92.

Turner, N., Barling, J. and Zacharatos, A. (2002) 'Positive psychology at work', in C.R. Snyder and S.J. Lopez (eds) *The Oxford handbook of positive psychology*, New York: Oxford University Press, pp 715-28.

Ulrich, D. (2010) 'Foreword. The abundant organization', in A. Linley, S. Harrington and N. Garcea (eds) *Oxford handbook of positive psychology and work*, Oxford: Oxford University Press, pp xvii-xxiii.

Ulrich, R. (1984) 'View through a window may influence recovery from surgery', *Science*, vol 224, no 420e1.

Uttal, W.R. (2001) *The new phrenology. The limits of localizing cognitive processes in the brain*, London: The MIT Press.

Uzzell, D. and Moser, G. (2009) 'Introduction: environmental psychology on the move', *Journal of Environmental Psychology*, vol 29, no 3, pp 307-8.

Uzzell, D. and Räthzel, N. (2009) 'Transforming environmental psychology', *Journal of Environmental Psychology*, vol 29, no 3, pp 340-50.

van Bavell, R., Herrmann, B., Esposito, G. and Proestakis, A. (2013) *Applying behavioural sciences to EU policy-making*, Luxembourg: Publications Office of the European Union.

van Oorschot, K., Haverkamp, B., van der Steen, M. and van Twist, M. (2013) *Choice architecture. Working paper of the NSOB*, The Hague: Nederlandse School voor Openbaar Bestuur (NSOB).

Vidal, F. (2009) 'Brainhood, anthropological figure of modernity', *History of the Human Sciences*, vol 22, no 1, pp 5-36.

Vul, E., Harris, C., Winkielman, P. and Pashler, H. (2009) 'Puzzlingly high correlations in fMRI studies of emotion, personality, and social cognition', *Perspectives on Psychological Science*, vol 4, no 3, pp 274-90.

Walkerdine, V., Lucey, H. and Melody, J. (2001) *Growing up girl: Psycho-social explorations of gender and class*, London: Palgrave.

Walmsley, D.J. and Lewis, G.J. (1993) *People and environment. Behavioural approaches in human geography* (2nd edn), Harlow: Longman Scientific and Technical.

Wastell, D. and White, S. (2012) 'Blinded by neuroscience: social policy, the family and the infant brain', *Families, Relationships and Societies*, vol 1, no 3, pp 397-414.

West, B.J., Patera, J.L. and Carsten, M.K. (2009) 'Team level positivity: investigating positive psychological capacities and team level outcomes', *Journal of Organizational Behavior*, vol 30, no 2, pp 249-67.

Wetherell, M. (2012) *Affect and emotion: A new social science understanding*, London: Sage Publications.

Wheeler, B. (2014) 'Whatever happened to the happiness agenda?', *BBC News*, 16 January (www.bbc.co.uk/news/uk-politics-25746811).

Wheeler, M. (2005) *Reconstructing the cognitive world: The next step*, Cambridge, MA: The MIT Press.

Wicker, A.W. (1979) *An introduction to ecological psychology*, Belmont, CA: Brooks/Cole Publishing Company.

Williams, M. and Penman, D. (2011) *Mindfulness: A practical guide to finding peace in a frantic world*, London: Piatkus.

Williamson, B. (2014) 'Knowing public services: cross-sector intermediaries and algorithmic governance in public sector reform', *Public Policy and Administration*, vol 29, no 4, pp 292-312.

Wolpert, J. (1964) 'The decision process in spatial context', *Annals of the Association of American Geographers*, vol 54, no 4, pp 537-58.

Wolraich, M.L., Feurer, I.D., Hannah, J.N., Baumgaertel, A. and Pinnock, T.Y. (1998) 'Obtaining systematic teacher reports of disruptive behavior disorders utilizing DSM-IV', *Journal of Abnormal Child Psychology*, vol 26, no 2, pp 141-52.

Wright, T.A. and Quick, J.C. (2009) 'The emerging positive agenda in organizations: greater than a trickle, but not yet a deluge', *Journal of Organizational Behavior*, vol 30, no 2, pp 147-59.

Wright, T.A., Cropanzano, R., Bonett, D.G. and Diamond, W.J. (2009) 'The role of employee psychological wellbeing in cardiovascular health: when the twain shall meet', *Journal of Organizational Behavior*, vol 30, no 2, pp 193-208.

Yen, J. (2010) 'Authorizing happiness: rhetorical demarcation of science and society in historical narratives of positive psychology', *Journal of Theoretical and Philosophical Psychology*, no 30, vol 2, pp 67-78.

Zeidner, M., Matthews, G. and Roberts, R.D. (2004) 'Emotional Intelligence in the workplace: a critical review', *Applied Psychology*, vol 53, no 3, pp 371-99.

Zeisel, J. (2006 [1981]) *Inquiry by design. Environment/behaviour/neuroscience in architecture. Interiors, landscape, and planning* (2nd edn), London: W.W. Norton & Co.

Zeisel, J., Silverstein, N.M., Hyde, J., Levkoff, S., Lawton, M.P. and Holmes, W. (2003) 'Environmental correlates to behavioral health outcomes in Alzheimer's special care units', *Gerontologist*, vol 43, no 5, pp 697-711.

Zerilli, L.M.G. (2013) 'Embodied knowing, judgment, and the limits of neurobiology', *Perspectives on Politics*, vol 11, no 2, pp 512-15.

Index